2022 Distinguished Instructor Short Course

Distinguished Instructor Series, No. 25

sponsored by the

Society of Exploration Geophysicists

Distributed Acoustic Sensing for Seismic Measurements – What Geophysicists and Engineers Need to Know

presented by

Mark E. Willis

SOCIETY OF EXPLORATION GEOPHYSICISTS

ISBN 978-1-56080-086-6 (Series)
ISBN 978-1-56080-384-3 (Volume)

Library of Congress Control Number: 2022942709

Society of Exploration Geophysicists
10300 Town Park Dr. Ste SE 1000
Houston, TX 77072 USA

Published 2022
Reprinted 2023, 2024

Printed in the United States of America

About the Author

Mark E. Willis is the Chief Scientific Advisor of Borehole Seismics at Halliburton. He is responsible for mentoring technologists, developing and promoting geophysical innovations, and fostering long-term client relationships. Previous to joining Halliburton in 2011, he worked in various research technology, supervisory, and management positions at Mobil Oil, Cambridge GeoSciences, Massachusetts Institute of Technology (MIT) Earth Resources Laboratory, and ConocoPhillips.

In his career, he has performed research and development in distributed acoustic sensing, vertical seismic profiling (VSP) technology, deep sonic log imaging, fracture identification using seismic data (time lapse VSP, microseismics, and surface seismic scattering), interferometric imaging, Kirchhoff and reverse time depth migration, full waveform inversion, machine learning, velocity model building, and sonic waveform processing.

Willis holds a B.S. in applied math and physics from the University of Wisconsin–Milwaukee and a Ph.D. in geophysics from MIT. He has written more than 100 papers, publications, and presentations, and he holds multiple patents. He is a member of SEG, SPE, EAGE, SSA, SPWLA, and ASA.

Contents

Preface

The secret of getting ahead is getting started.

– Mark Twain

Enter these enchanted woods, you who dare.

– George Meredith

My goal in writing this book and making the SEG DISC class presentations is to demystify distributed acoustic sensing (DAS). These past 10 years have been a heady time – we have seen many advancements of the interrogator hardware, the methods to process the data, and our general understanding of the technology. My first real introduction to DAS was in 2013 after Halliburton acquired SensorTran in 2011, and we performed our first wireline DAS VSP acquisition at the University of Texas at Austin's Devine Test Site, located in Devine, Texas. We had a team of geophysicists, optical engineers, field engineers, and software developers who brought this technology through its growing pains and finally into a viable FiberVSP™ service.

When I started, there were many questions about the technology. These included:

1. Fiber is not a geophone, so how does it actually work?
2. How do we compare the data we obtain from DAS to our conventional seismic data?
3. What is all of that strange noise we see on DAS records?
4. How can we increase the signal-to-noise ratio (S/N) of the DAS data?
5. Can we trust the amplitudes we obtain from DAS data?
6. How do we choose the correct hardware and settings to obtain the best data?

I am very happy to report that we have answers to all of these questions. I've endeavored to provide those answers in this book in a format that will be easy to follow and understand. I've tried to include as many references as practical for all of the different aspects of the technology. If you have questions that I have left unanswered, hopefully I have pointed you to at least some of the original sources of materials, knowing that in my database of DAS papers there are about 1000 references out of probably three-fold as many available.

I spent the first year of preparation for this book and class creating an open source MATLAB application, ExploreDAS, that will allow you learn about DAS by "doing." I thought it would allow me to create a large portion of the illustrations for the book. However, I realized later that I would still need to create the teaching materials with one-off examples. I hope that the ExploreDAS software will allow you on your own to test ideas and discover how DAS works.

Acknowledgments

Alone we can do so little; Together we can do so much.

– Helen Keller

No passion in the world is equal to the passion to alter someone else's draft.

– H. G. Wells

I thank my wife, Lynn Willis, for her encouragement and support throughout this project.

Organizations: I thank SEG for selecting me to be a part of the long line of previous excellent DISC instructors. I hope to live up to those standards. I also thank Halliburton for allowing me to participate in this project. A special thanks to Mathworks for providing free access to MATLAB licenses for all of the students while attending the DISC class. I very much appreciate the permission to use copyrighted figures from EAGE publications in this book.

Optical professionals: I am forever indebted to David Barfoot who introduced me to DAS technology and patiently answered my naïve questions. I also have learned so much from other team members: Andreas Ellmauthaler, Michel LeBlanc, John Maida, Etienne Samson, Jeff Bush, and Glenn Wilson.

Geophysical professionals: I have greatly enjoyed working and learning alongside Xiang Wu, Amit Padhi, Oscar Barrios Lopez, Pedro William Palacios, Cemal Erdemir, Ran Zhou, George Knapo, Laura Torres, Paul Peterman, Janusz Peron, Jose David Carrillo Rangel, Dongjie Cheng, Michelle Xiaomin Zhao, Kary Green, Minyu Zhang, Sonali Pattnaik, Kris Walker, Victor Leung, and Jason Gibbs.

Software professionals: Special thanks to the software team which included Chen Qi, Arun Kumar, Rongrong (Rory) Lin, Ihor Lapko, and Mendel Vladyslov, as well as many others.

Field operations: Without great field support, this is all for nothing. I have greatly benefited from the work and collaboration of Simon Shaw, Henry Bland, and Jason Duplechain.

Management and project management: Over this time period, I've been a member of the DAS team at Halliburton which has had extraordinary support from Dan Quinn, Avinash Taware, Arabinda Mishra, Arthur Cheng, John Granville, Robert Gales, Milos Milosovic, Eliot Rowden, Wayne Tu, and Kristie Cheng.

Industry and academic colleagues: We live in an interconnected world, and I am continually inspired by my "optical" colleagues, including Sudhish Kumar Bakku, Albena Mateeva, Jorge Lopez, Samantha Grandi, Jonathan Ajo-Franklin, Ali Tura, Ge Jin, Yingping Li, Martin Karrenbach, Robert R. Stewart, Mahmoud Farhadiroushan, Elita Yunyue Li, Jingjing Zhong, Zohreh Souri, Tanya Inks, Aleksei Titov, Gary Binder, Dana Jurick, Erkan Ay, Ezzedeen Alfataierge, Andrey Bakulin, Arthur Hartog, Baishali Roy, Eileen Martin, Andreas Chavarria, Timothy Dean, Roman Pevzner, and Ge Zhang, among many others.

Reviewing and editing: I heartily thank John Maida, Oscar Augusto Barrios Lopez, Robert Gales, Kiattisak James Petpisit, Andreas Ellmauthaler, Henry Bland, Amit Padhi, Michel LeBlanc, and Xiang Wu for carefully reviewing the manuscript, providing edits, and helping me resolve some of the finer issues. I also thank Melissa Presson, Susan Stamm, and Jennifer Cobb from SEG for helping me through the process of creating the DISC course and this book.

Chapter 1 Applications of fiber optic distributed acoustic sensing technology

In the right light, at the right time, everything is extraordinary.

– Aaron Rose

1. Application of seismic methods overview

It is my great pleasure to introduce you to some of the amazing aspects of fiber optic distributed acoustic sensing (DAS) technology. In order to motivate the need to better understand this technology, I'll start by reviewing some of the diverse applications that are using it. You may be familiar with some of these applications but perhaps not all of them. I'll delve into an explanation of DAS physics and instrumentation in subsequent chapters.

Just for fun, you can take a short quiz to see how much you already might know about DAS technology using a MATLAB gui (Willis, 2022a).

1.1. Earthquake monitoring

We need to record transient seismic waves propagating in the earth for many reasons. From historical time to the present, humankind has feared earthquakes because they can inflict catastrophic damage on buildings, infrastructure, and the landscape. This has driven the development of measurement devices to record and characterize the ground motion from them. From a single sensor seismoscope system invented in AD 132 by Zhang Heng (Mitrica, 2019) to current networks of seismometers such as the USArray (USAarray.org; Bragato et al., 2021; USGS.gov) capable of detecting and recording earthquakes, humankind greatly values the ability to learn about the forces creating these events.

Earthquakes are the result of rocks breaking and yielding due to tectonic stresses that exceed their strength. At the global scale, the boundaries of the earth's tectonic plates are in constant stress due to the dynamic forces driving them together, pulling them apart, or slipping past each other by shearing motion. These enormous stresses have the ability to induce large-scale earthquakes. On a smaller scale, local stresses also cause earthquakes. The rock breaking from each of these earthquakes generates a cafeteria full of body and surface waves that propagate outward from the event. Figure 1.1 shows a simulation of the wavefield that has propagated inside the earth from an earthquake.

By measuring and recording all of these earthquakes, seismologists have been able to interpret the travel paths of these waves to determine the structure of the deep and shallow earth. Figure 1.2 is an illustration of the ray paths some of the many seismic waves take through the earth.

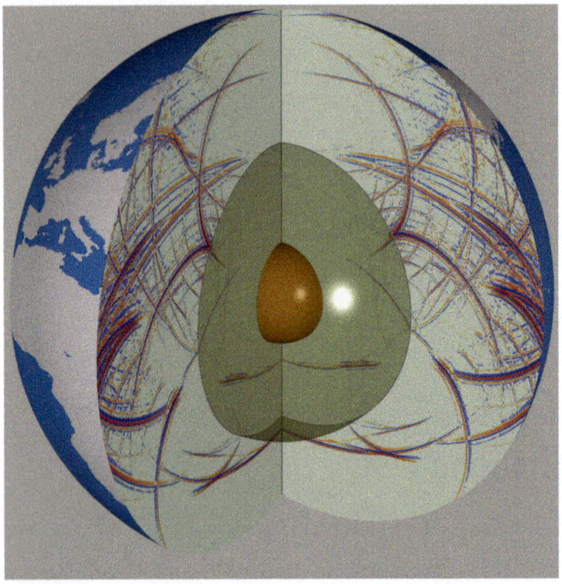

Figure 1.1. A visualization of the simulated wavefield of seismic energy created from an earthquake. (Courtesy Heiner Igel, https://seismology.geophysik.uni-muenchen.de/images/earth_big.jpg.)

You can view earthquake records in near real time from Iris at https://www.iris.edu/app/station_monitor/. Figure 1.3 is an example of an earthquake recorded at Hockley Texas Station (HKT_IU) using three-component seismometers.

While the primary measurement device for earthquakes is currently seismometers, DAS sensors are being used in some applications especially for analyzing microseismic events and surface waves. Figure 1.4a shows an example of a magnitude 2.6 earthquake recorded by a fiber optic DAS system. In the figure, a 100 km long fiber is recorded for 3 seconds, where increasing seismic amplitude is plotted as a color shift from green to red. The P-waves are the first event visible (marked with the green line) and S-waves are the second event (marked by the blue line). The DAS systems inherently measure only a single directional component of a wavefield, which makes it more difficult to identify which arrival is the P-wave and which arrival is the S-wave.

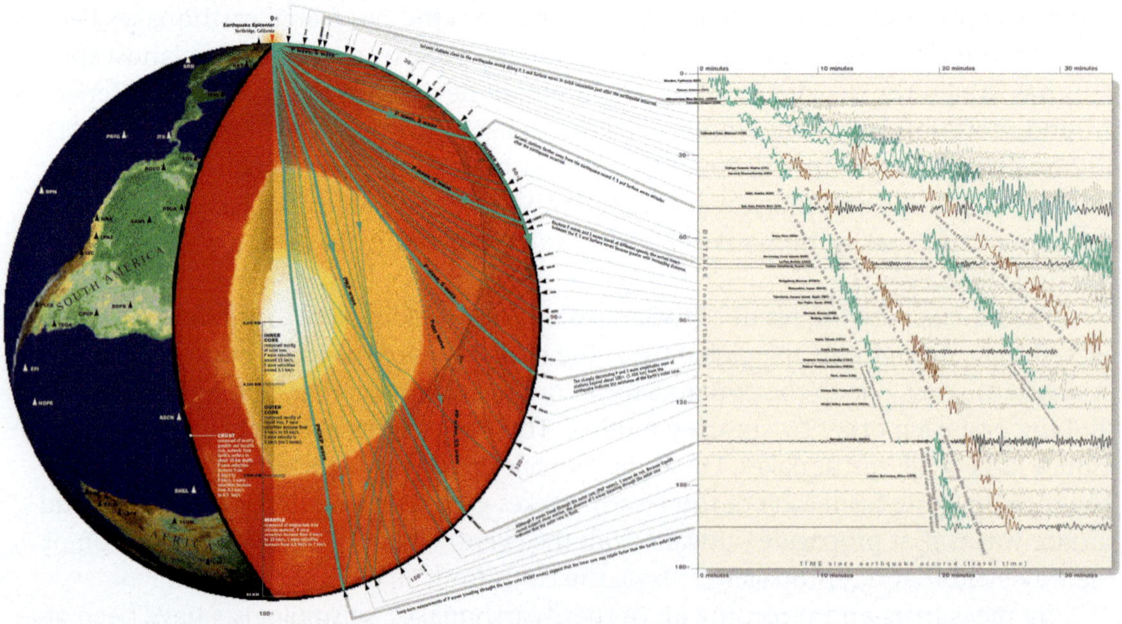

Figure 1.2. Cross section earth view (left) showing ray paths for a collection of seismic waves in the earth. The corresponding seismograms (right) are shown for some of these events (https://www.iris.edu/gallery3/general/posters/exploring_earth/WavePropagation).

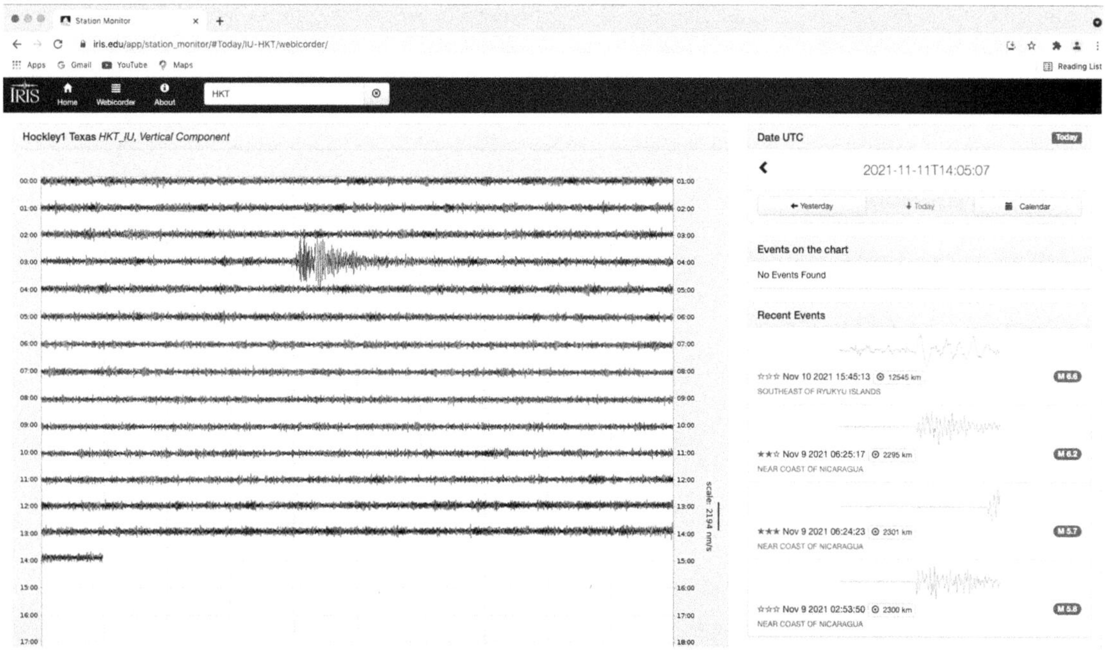

Figure 1.3. Example earthquake record using the IRIS website for the Hockley Texas Station.

Lindsay et al. (2020) investigates the broadband instrument response of DAS systems comparing them with conventional seismometers used for teleseismic events and finds excellent agreement when the DAS data is converted to the seismic response. Figure 1.4b shows the Guralp CMG-3T seismometer recording (top row) and its spectrogram (second row) for the 23 January 2018 Alaska magnitude 7.9 event. The corresponding DAS recording (third row) and spectrogram (fourth row) show similar waveforms and spectral content which confirms its suitability for use recording teleseismic events. Ajo-Franklin et al. (2019) describe using dark fiber (unused telecommunications fiber optic cables) for earthquake seismology extracting shallow velocity profiles and regional and teleseismic earthquakes. Figure 1.4c shows a comparison of the Rayleigh phase arrivals from a variety of teleseismic events at the same station/array pair. The inset shows the station location (arrow) and the DAS array geometry. These examples show the near-perfect match between the seismometer and DAS recordings.

Society is greatly concerned about understanding earthquakes and determining how to avoid destruction from these events. Figure 1.5 shows damaged buildings from the magnitude 6.6 April 20, 2013 Lushan Earthquake in China. Figure 1.6 shows damaged cars and buildings from a 1994 magnitude 6.7 earthquake in the Northridge area of Los Angeles, California, which left 51 people dead and 7,000 injured. Figures 1.7 and 1.8 show damaged bridges from the Northridge and Loma Prieta earthquakes in California.

We also monitor earthquakes to determine the risks from volcanoes. The seismic events from volcanoes can help understand how magma and gases move within the crust. Being able to distinguish between volcano-tectonic, long period, and tremor events is key to predicting whether an eruption is imminent. Figure 1.9 is a diagram showing the many technologies used to monitor volcanoes, including seismometers.

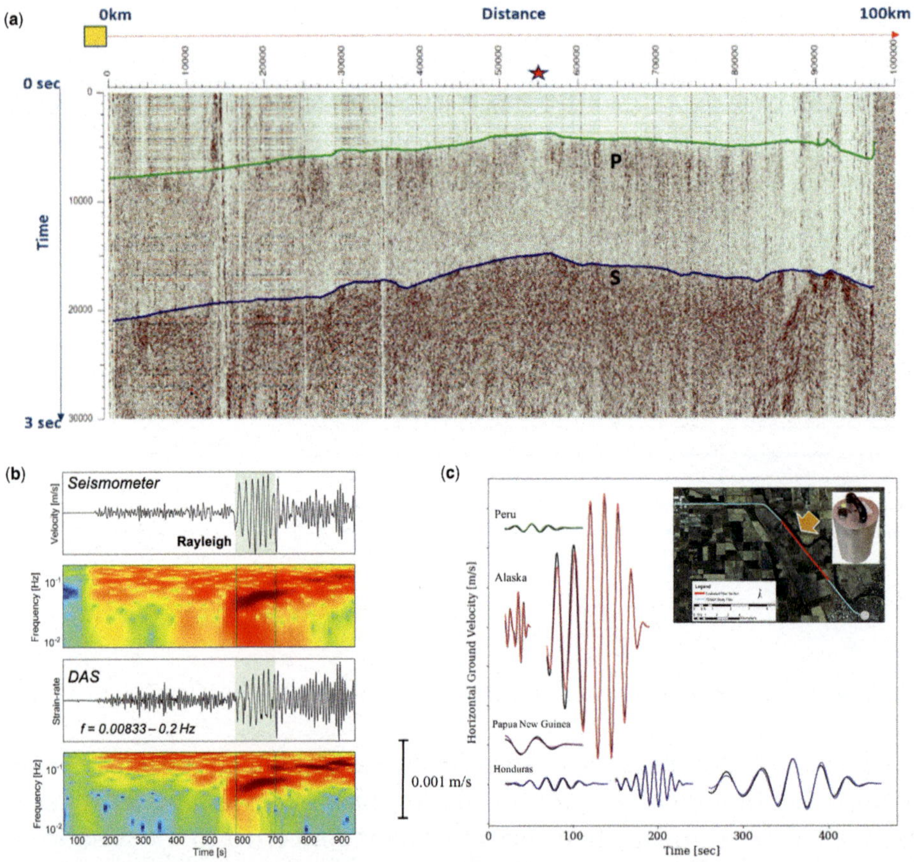

Figure 1.4. (a) A magnitude 2.6 earthquake record created by a fiber optic DAS system. The P- and S-waves are denoted by the green and blue lines, respectively. (After Figure 7 of Karrenbach et al., 2021.) (b) Recordings of the 23 January Alaska magnitude 7.9 earthquake. First row: seismometer record; second row: spectrogram of the seismometer record; third row: DAS record; and fourth row: spectrogram of the DAS record. (Personal communication, courtesy of J. Ajo-Franklin and N. Lindsey.) (c) Comparison of the Rayleigh phase arrivals for a variety of teleseismic events. (Personal communication, courtesy of J. Ajo-Franklin and N. Lindsey.)

Figure 1.5. Damaged buildings from the April 20, 2013 earthquake in Lushan, China. (Courtesy USGS.)

Figure 1.10 is a photo of two scientists maintaining a volcanic seismic station.

1.2. Facilities monitoring

Other applications of seismic data acquisition have included perimeter intrusion detection around important infrastructures, roads, and buildings using sensors deployed in proximity to these structures. In fact, Jean Perrin invented the geophone in 1915 for WWI troop detection (Sloan et al., 2021).

Figure 1.6. Damaged cars and buildings from the 1994 Northridge, CA earthquake. (Courtesy USGS.)

Figure 1.7. Damaged bridge from the 1994 Northridge earthquake. (Courtesy USGS.)

In urban areas, seismic technology in the form of fiber optic distributed acoustic sensing can be used to detect foot and vehicle traffic on private property. Here, cables are buried in shallow trenches or attached to fences and buildings so that the ground motion from intruders is recorded and analyzed to detect and locate them. Figure 1.11 shows a portion of a campus equipped with one of these systems, which is nearly impossible to visually detect.

Vibration can be monitored to characterize the flow of gases and liquids in pipelines and to assess the structural integrity of bridges and dams. Figure 1.12 is a photo of a water pipeline traversing the Iceland countryside. Pipelines such as this can be monitored with fiber optic DAS sensing to locate possible leaks or damage from animals, corrosion, shifting foundations, and physical attacks.

Figure 1.8. Road damage from the 1989 Loma Prieta, CA earthquake. (Courtesy USGS.)

There are many scholarly papers on using seismic monitoring with DAS measurements for characterizing road and sidewalk pedestrian traffic. In urban areas, there are usually many sets of fiber optic cables enclosed in underground conduits to provide telecommunications to area homes and businesses. Some of these fibers are unused, referred to as "dark fiber," so they may be available for use in detection systems. Arrangements can be made with utility service providers to access these dark fibers and

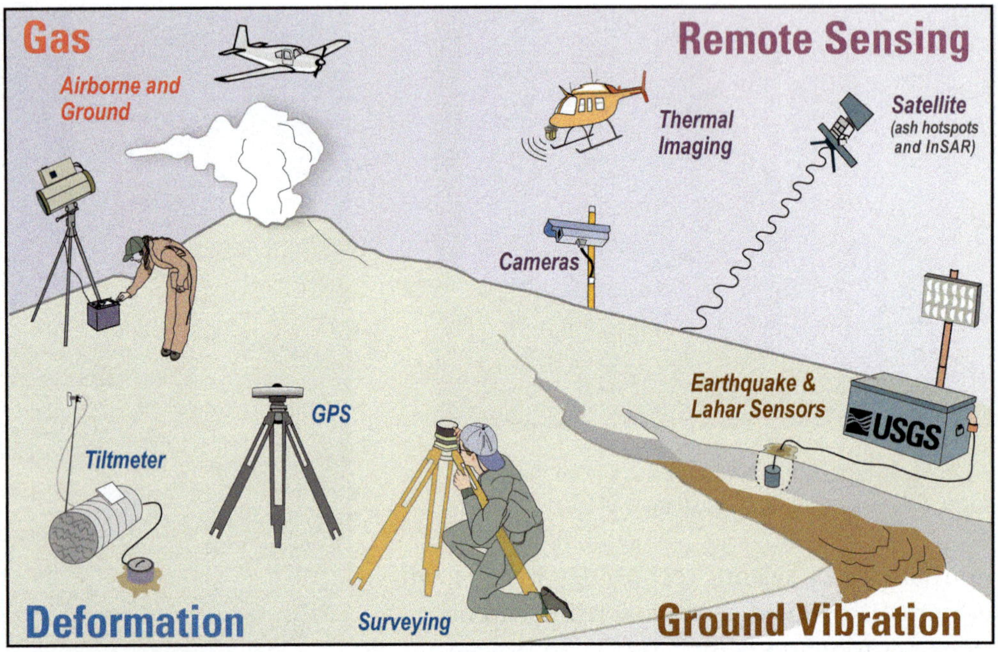

Figure 1.9. Diagram showing the many technologies used to monitor volcanoes. (Courtesy Lisa Faust, public domain; https://www.usgs.gov/media/images/volcanic-monitoring-types-and-methods-employed-usgs-vhp.)

Figure 1.10. Two scientists maintaining a volcano seismic station. (Courtesy Chris Waythomas, public domain; https://www.usgs.gov/media/images/station-maintenance-veniaminof-volcano-ak.)

instrument them with DAS interrogator systems so that acoustic measurements of cultural and environmental signals can be recorded.

Figure 1.13a shows a map of an area on the Penn State University campus where researchers obtained access to dark fiber and made DAS measurements. Figure 1.13b shows a record covering approximately 4 minutes during which a bus stops at time 9:54:30. The bus lets out seven people who walk south starting at DAS channel 1391 until about channel 1451 at a time of about 9:56:04. Then, at time 9:56:30, a person begins to walk in the opposite direction back toward channel 1931. This example demonstrates the incredible accuracy and sensitivity that DAS has to measure sound, even by fibers which were designed for other applications and located in underground bunkers.

Buried fiber also can be used to monitor road traffic. Figure 1.14a shows DAS recordings made next to a highway. The multicolored lines in Figure 1.14b show the location of the fiber optic cable whose position zigzags along its generally NE/SW path. It traverses that path back and forth three times. A two-lane highway is to the west of the fiber optic cable and a service road passes parallel through the middle of the path of the fiber. In Figure 1.14a, the vertical axis is the channel number and the horizontal axis is time which spans 30 minutes. The bright red and yellow lineations show ground motion from cars entering the fiber line from the NE at about channel 3100 and moving SE down the road toward channels 0 and 5500. The color coding in Figure 1.14b is the strain amplitude of energy in

Figure 1.11. A forest setting around a campus which has been equipped with buried fiber optic DAS equipment to identify intruders. (Courtesy Senstar.)

Figure 1.12. Pipelines require protection from attacks. (Photo by Mike Benna on Unsplash, https://unsplash.com/photos/X-NAMq6uP3Q.)

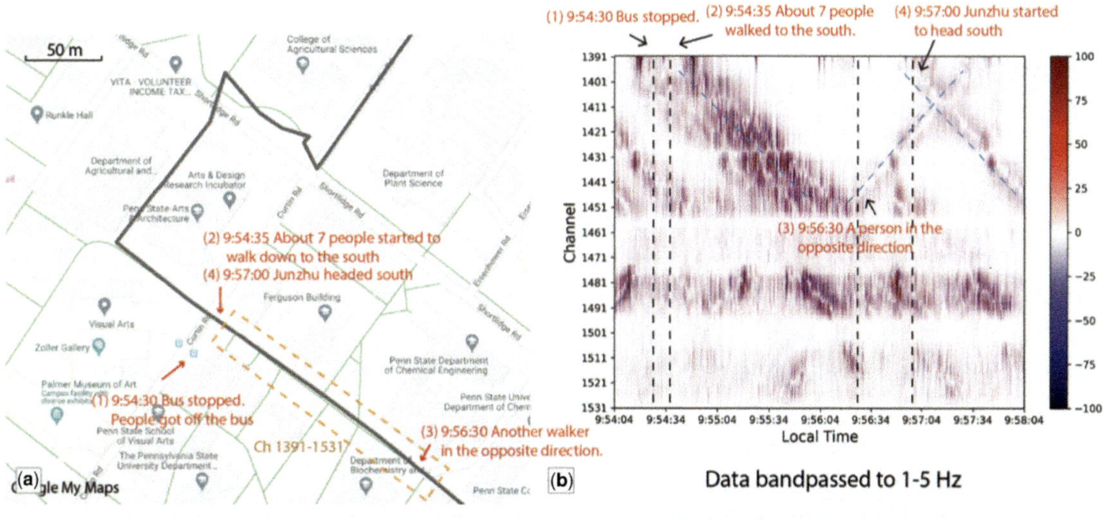

Figure 1.13. (a) Map of a portion of Penn State University campus with underground fiber optic communication cables installed. The map shows the location of a bus that stopped and released passengers who walked along the sidewalk above the buried fiber cables. (b) The DAS recording of the acoustic signal from pedestrians walking on the sidewalk above the buried fiber cable. (Modified from Figure 3 of Jakkampudi et al., 2020.)

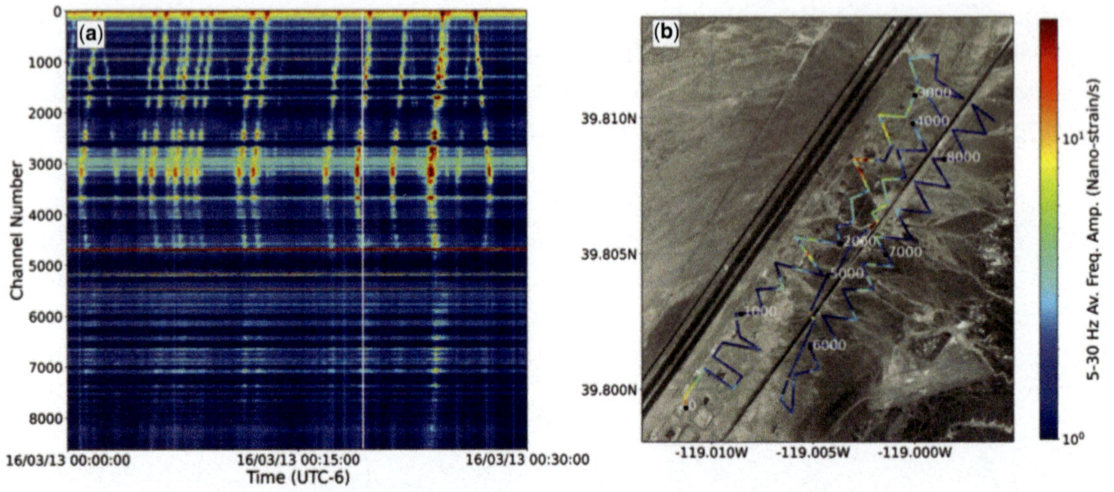

Figure 1.14. Fiber optic DAS measurements made near a highway. (a) The DAS strain rate recording with the vertical axis indicating the channel number and the horizontal axis indicating approximately 15 minutes of time. (b) An aerial photo showing the highway (on the bottom left to upper right) and the service road to the east. The color zigzag line shows the location of the fiber optic cable. The colors indicate the strain rate amplitudes for a time window indicated by the white line in (a). (From Figure 4 of Chambers, 2020.)

one particular time slice (shown by the vertical white line in Figure 1.14a at about time 00:18:00) when a vehicle was located at about channel 2600.

The changing arctic climate conditions have forced polar bears to search for food outside of their normal habitat. In order to protect arctic science stations in Greenland, the U.S. National Science Foundation, Office of Polar Programs, Arctic Sciences, Research Support and Logistics Program (NSF-OPP-ARC-RSL) has sponsored research to determine the effectiveness of fiber optic DAS intrusion detection systems to monitor facilities for "visiting" polar bears (Glaser et al., 2021). In an instrumented cold snow test cell room, they have simulated the quadruped locomotion characteristics of polar bears in contrast to human bipedal characteristics. In addition, they have considered factors which might make the system fail, such as potentially poor fiber coupling in snow and the effect of very cold temperatures (–70°C) on the sensitivity of the fiber to acoustic signals.

Not only can seismic measurements be used simply to detect and locate vehicles and people, but the signal itself can be used to characterize the near surface earth conditions. Figure 1.15a shows the locations of a high-density seismic network of geophones in the Long Beach, CA area. Figure 1.15b shows a map of the area with local roads and waterways. Using traffic noise only, Chang et al. (2016) use tomography inversion of the recorded surface (Rayleigh waves) to extract the velocity profile of the near subsurface. Figure 1.15c and 1.15d shows different views of the velocity profile of the near subsurface.

Fiber optic DAS systems also are being used to characterize the near surface structure. Ajo-Franklin et al. (2019) describe the use of dark fiber in underground telecommunication systems to invert recordings to extract near surface structures using ambient noise interferometry. Fang et al. (2018) use blasts from a quarry located approximately 13 km away and dark fiber fiber optic DAS seismic data to analyze the

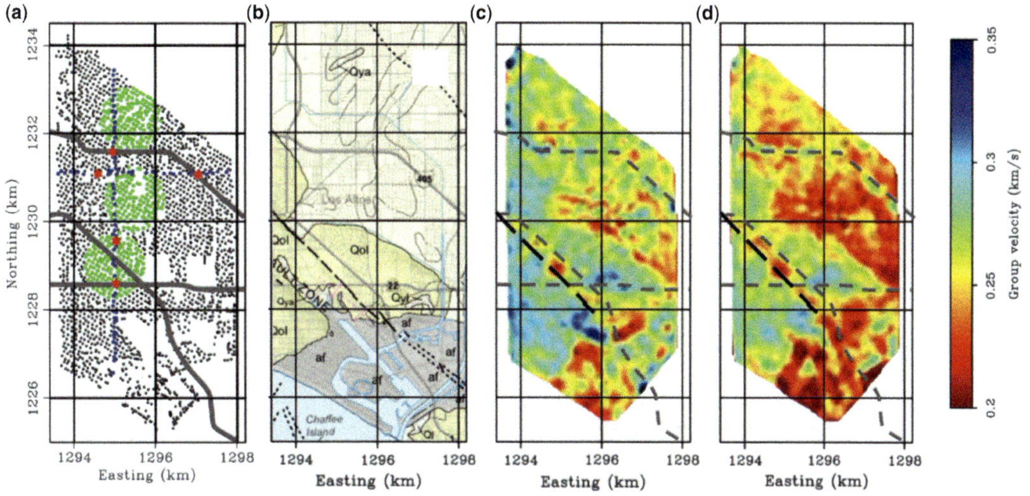

Figure 1.15. (a) Map of the location of geophone sensors in the Long Beach, CA area. (b) Map of the Long Beach, CA area. Group velocity map of the near surface using (c) 3 Hz and (d) 3.5 Hz bandwidth. (Modified from Figures 1 and 9 of Chang et al., 2016.)

change in the near surface due to the construction of a basement in a building on the Stanford University campus.

1.3. Exploration and production applications

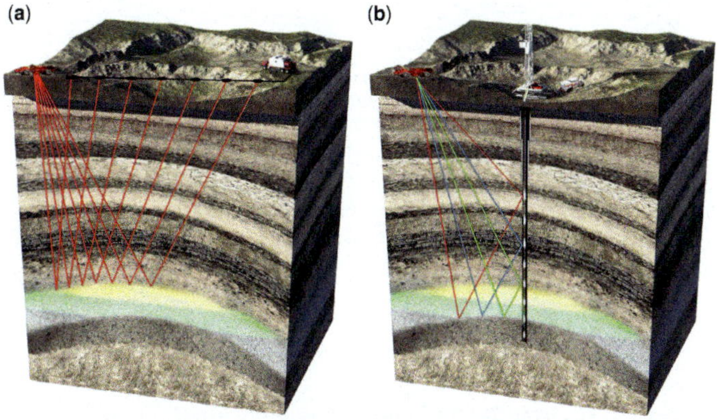

Figure 1.16. (a) Geometry of surface seismic data acquisition. (b) Geometry of VSP data acquisition.

In the oil and gas sector, there are several acquisition geometries used to acquire seismic data. The first geometry, called the surface seismic method, has the seismic sensors and sources on, or very near, the surface of the earth (Figure 1.16a). The second geometry, called vertical seismic profiling (VSP), has the sensors placed in a borehole and the sources are on the surface of the earth (Figure 1.16b).

For the land surface seismic geophone case, geophones are planted in the soil. Alternatively, the geophones can be buried in shallow holes. For the land DAS case, fiber optic cables are placed on the surface of the ground. However, the coupling tends to be very poor for cables simply laid on the ground. To address that issue, shallow trenches can be dug and the fiber optic DAS cable can be laid in the bottom and covered with soil.

In both the geophone and DAS cases, a surface source is used to create seismic energy which propagates down into the earth and is reflected back to the sensors on the surface. These reflected waves when processed allow images to be made of the subsurface. The seismic source can be one of many options including: an explosive source like dynamite, a vibratory source like a vibroseis truck, or weight drops. Figure

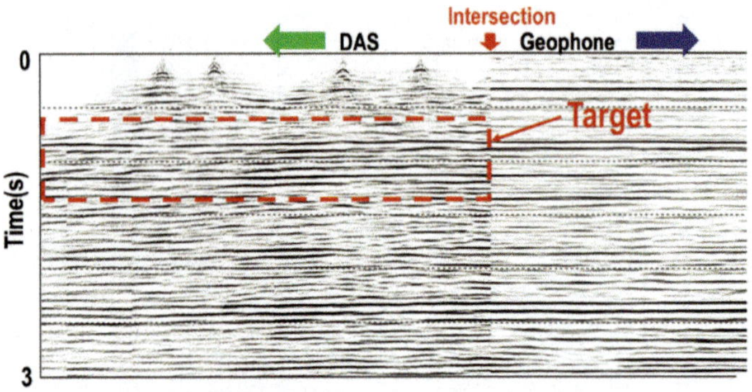

Figure 1.17. Comparison of surface seismic images from DAS (left) and geophones (right). (Modified from Figure 3 of Bakulin et al., 2020.)

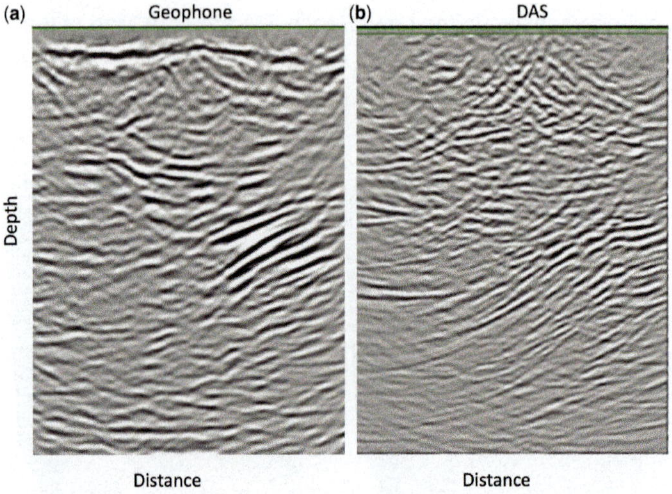

Figure 1.18. Example of surface seismic imaging using (a) geophones on the surface and (b) DAS fiber sensing on the surface. (Modified from Figure 4 of Bona et al., 2021.)

1.17 shows a comparison of images of the subsurface created using DAS fiber optic cable on the left and geophones on the right. The value of using DAS is the high density of receiver locations when compared to geophones, which has the potential of creating higher resolution images.

Figure 1.18 shows another example comparing surface seismic data acquired with geophones (left) and DAS fiber optic cables (right). The DAS image in this case shows a remarkably higher detail of the structure especially on the middle right side of the image. There are notable differences in the images, such as the lack of the very shallow event in the DAS image and the striking differences in the possible structural interpretations. These may be due, in part, to the response of the fiber, the density of receiver locations, and processing.

Seismic sensors, either geophones or fiber optic cable, can be inserted in a borehole so that seismic data are collected from the perspective of the well. Data acquired using this geometry is called vertical seismic profiling (VSP). The goal of creating a VSP image is frequently several fold: 1) to increase the resolution of the subsurface image, 2) to tie the surface seismic image more accurately to depth, and 3) to remove multiple-reflected events from the image.

Figure 1.19a shows a migrated image created from surface seismic geophone data. Figure 1.19b shows the image created from VSP using DAS data inserted into the middle of the surface seismic image. You can clearly see that the events in the VSP data seem to be more continuous and sharper.

A variation of the surface seismic method is the ocean bottom node (OBN) method. This geometry makes use of self-contained recording units placed on the ocean bottom. These units contain three component geophones and one hydrophone with an independent recording system. These units are placed on the ocean bottom by robots at the beginning of the survey. A source boat with air gun seismic sources travels over the

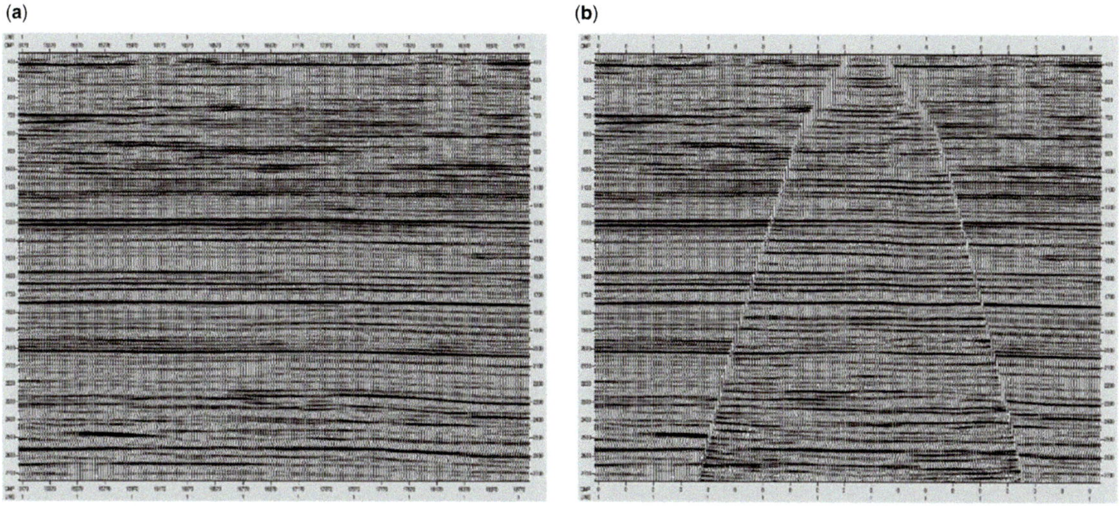

Figure 1.19. Comparison of (a) surface seismic image, and (b) DAS VSP image superimposed on (a). (Figure 4 of Duan et al., 2021.)

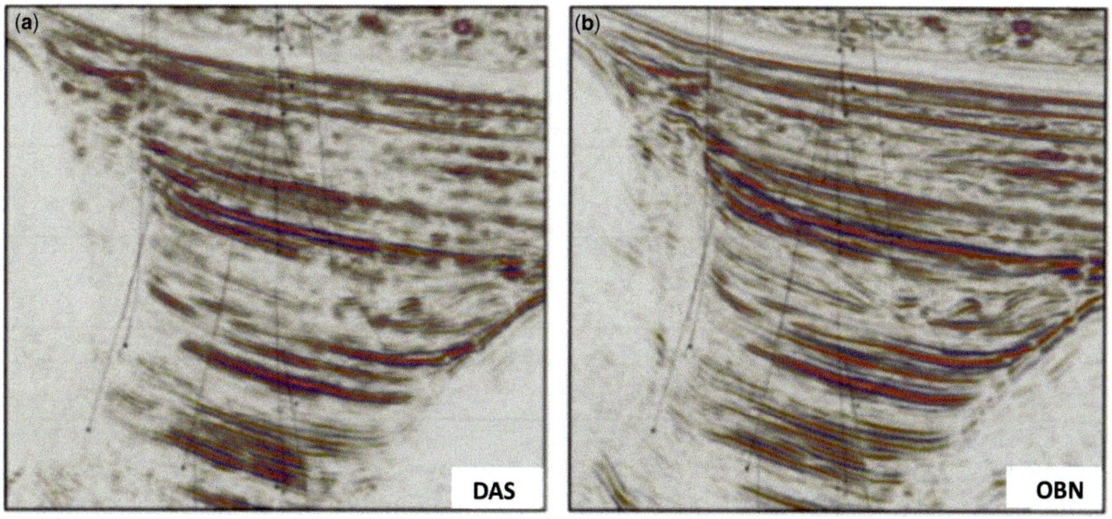

Figure 1.20. Multi-well (a) DAS VSP image and (b) corresponding OBN image. (Modified from Figure 7 of Zhan and Nahm, 2020.)

survey area while the OBN units record all the shots. Later, the OBN units are retrieved and the data are processed to form images of the reservoir.

Figure 1.20a shows a migrated image of the Mars Field in the Gulf of Mexico using DAS VSP data acquired simultaneously in several wells. Figure 1.20b shows the same image created from OBN data. The images are structurally quite similar but also show some significant amplitude differences most likely caused by the limited aperture imposed by the VSP acquisition geometry. Because the OBN receivers are located in a wide area over the sea bottom and the VSP only has sensors in a limited number of wells, the lateral coverage is not the same. However, the great benefit is that the DAS VSP survey can be acquired with much less cost and time because the OBN receivers do not have to be mobilized and retrieved.

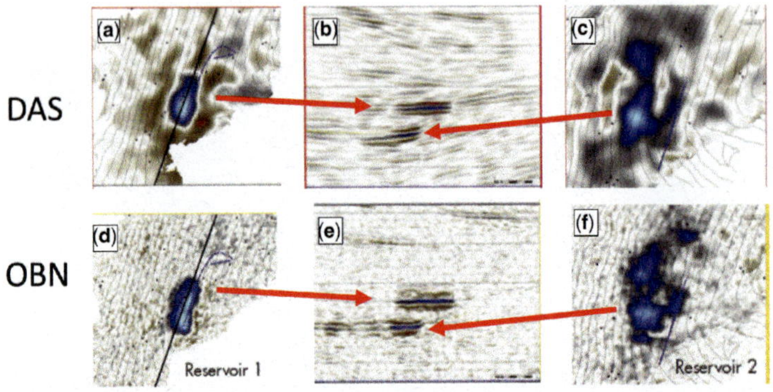

Figure 1.21. The 4D images of (a) VSP depth slice through upper reservoir, (b) selected VSP profile, (c) VSP depth slice through lower reservoir, (d) OBN depth slice through upper reservoir, (e) selected OBN profile, and (f) OBN depth slice through lower reservoir. (Modified from Figure 5 of Kiyashchenko et al., 2020.)

Typically, OBN data sets produce high-quality images which often are used in time lapse studies to determine the effective drainage pattern of a reservoir. The downside is that they are typically very expensive to acquire. A time lapse, or 4D, survey is performed by repeating a seismic acquisition of the same area separated by some time interval. In order to make changes in the production to optimize drainage of the reservoir, repeat seismic surveys should be made as often as possible. However, economics dictate that these surveys be made perhaps every six months or every year or two (Kiyashchenko et al., 2020).

Figure 1.21 shows a set of displays comparing 4D surveys using DAS VSP (top row) and OBN (bottom row). The center column shows the 4D survey images in cross section view. Figure 1.21b shows the DAS VSP 4D image, while Figure 1.21e shows the 4D OBN image. The OBN image has very clear, well resolved, isolated events in the upper (1) and lower (2) reservoirs. The DAS VSP image also has prominent time lapse effects in both reservoirs, and there is additional smaller, but noticeable, time lapse effects outside of the two reservoirs.

Figure 1.21a and 1.21d shows the map view through the upper (1) reservoir for the DAS and OBN data sets, respectively. Similarly, Figure 1.21c and 1.21f shows the map view through the lower (2) reservoir. Due to the limited aperture of the VSP data acquisition compared to the OBN data, there are many differences beyond the changes in the reservoir. However, the extent of the reservoir changes is easily discerned from the DAS VSP images which were acquired with much less cost than the OBN data set.

1.4. Passive seismic applications: Microseismic monitoring, low-frequency strain, and flow noise

The previous cases have used active seismic sources. When we only listen for seismic energy without actively introducing seismic sources under our control, we call it passive seismic. The three types I will cover are microseismic monitoring, low-frequency strain, and flow noise.

Whenever hydraulic stimulation is performed, high-pressure fluids are pumped into the well for the expressed purpose of creating new pathways for fluids to travel in the formation. This can be done for geothermal reservoirs, conventional reservoirs, and also for unconventional oil and gas reservoirs. When the rocks break under the pressure, microseismic events are created which can be recorded by sensors in a well or on the surface. Traditionally, three-component geophones in the well are used to monitor the

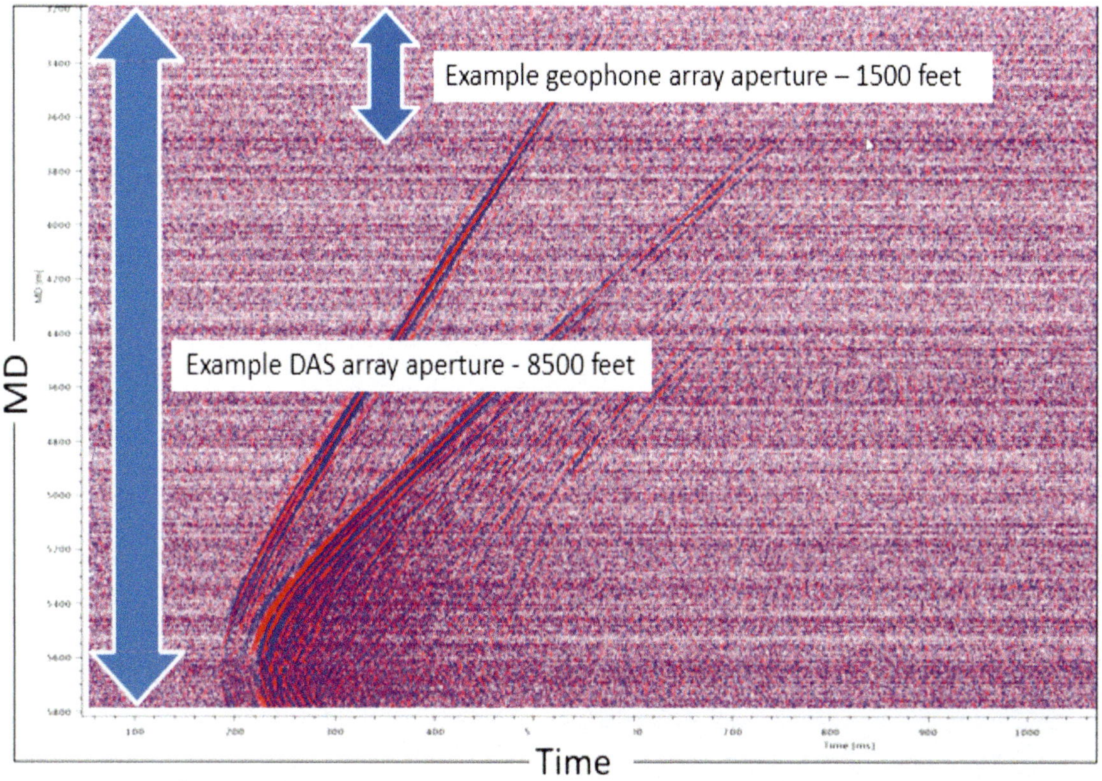

Figure 1.22. A DAS microseismic record spanning 8500 ft in depth and approximately 1 second in time. (Courtesy Halliburton.)

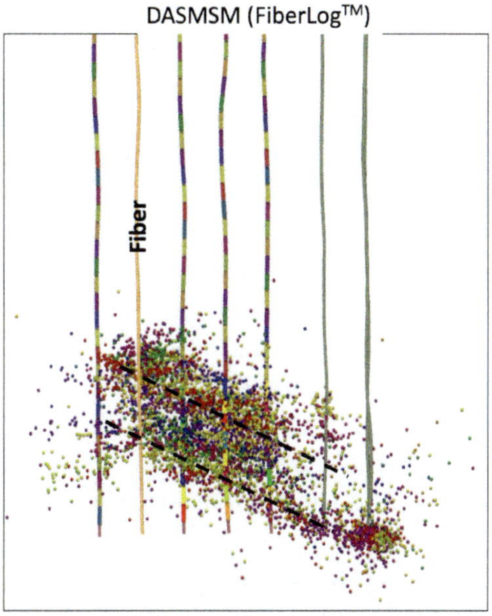

Figure 1.23. Map view of microseismic event locations (dots) shown with the wells being hydraulically stimulated. The yellow well contains the fiber optic cable. (Courtesy Halliburton.)

microseismic events because they easily allow the characterization and separation of P- and S-waves. For cost reasons, only a limited number of geophones can be placed downhole. In contrast, when using fiber optic DAS technology, greater than one order of magnitude more channels of data can be recorded (Willis et al., 2014).

Figure 1.22 shows an example microseismic event recorded by a fiber optic DAS acquisition system. Note that the extent of the recorded data is approximately 2600 channels spanning a measured depth of 8500 ft (shown by the large blue arrow). A typical geophone array might have only 20 channels over a depth range of approximately 1500 ft (shown by the small blue arrow).

Processing these microseismic records to detect the P- and S-wave arrivals, the location of each event can be determined (Hull et al., 2017). Figure 1.23 shows the map view of

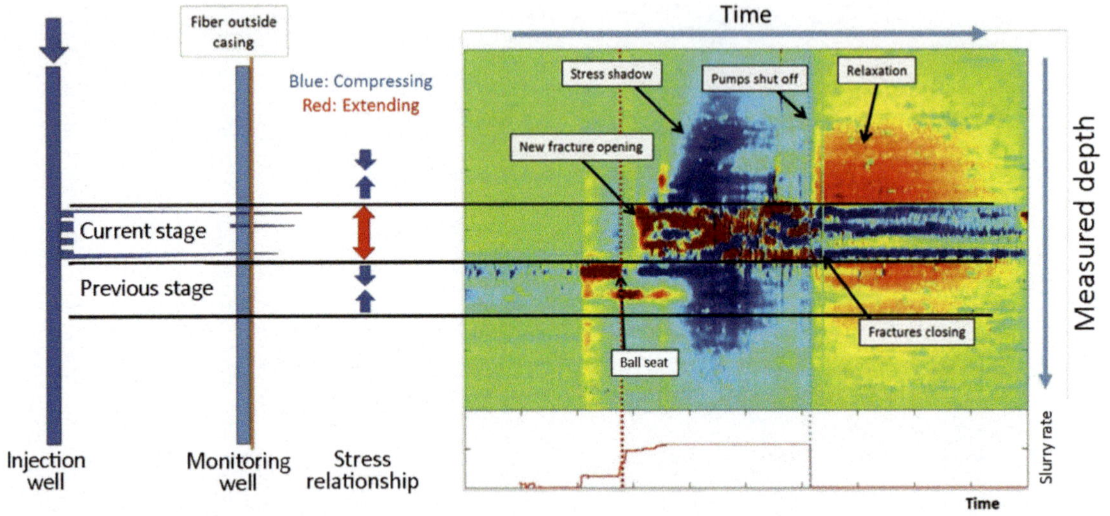

Figure 1.24. Low-frequency DAS response in a monitor well parallel to the injection well. (Modified from Figure 3 of Jin and Roy, 2017.)

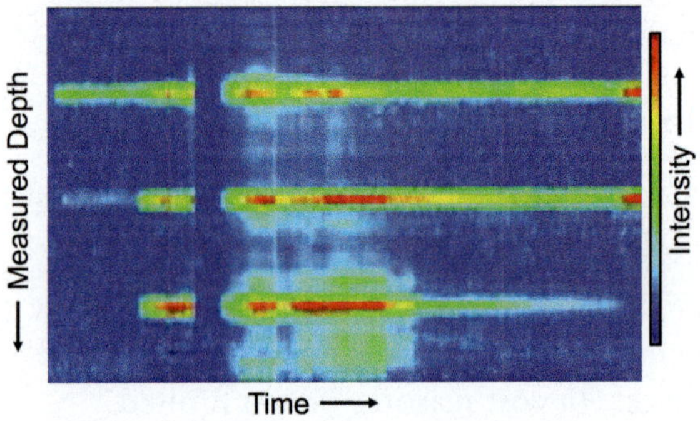

Figure 1.25. The DAS flow record showing acoustic noise as a function of measured depth (vertically) and time (horizontally). Flow from three perforated depths show high intensity. (Courtesy Halliburton.)

microseismic events which have been located from data recorded by DAS fiber. Microseismic events recorded by DAS only have one component (unlike geophones with three components), so assumptions have to be made in order to map the events to reasonable heuristic locations given that they are not as well constrained.

In addition to inducing microseismic events, the hydraulic fracturing introduces a low temporal and spatial frequency stress field into the subsurface that radiates away from the injection well. The result of this stress field is strain in the rocks that follows the creation of the fractures. Jin and Roy (2017) present seminal work demonstrating that the strain measured with a DAS fiber in a nearby lateral monitor well accurately captures the effects of the injected fluids as they propagate from the injection well toward the monitor well. In many cases, it seems that individual fractures emanating from each perforation in a cluster group can be identified and the relative success of them can be judged.

Another powerful use of DAS acoustic measurements is to monitor downhole flow noise. Whenever fluid moves in the borehole, it generates acoustic signals due to turbulence. Figure 1.25 shows an example of a waterfall plot of DAS acoustic energy for a depth interval (vertically) of a well for a time period (horizontally). The color indicates the noise intensity of the DAS measurement with red the most and blue the least energetic. Three clear perforations are generating a significant amount of turbulence.

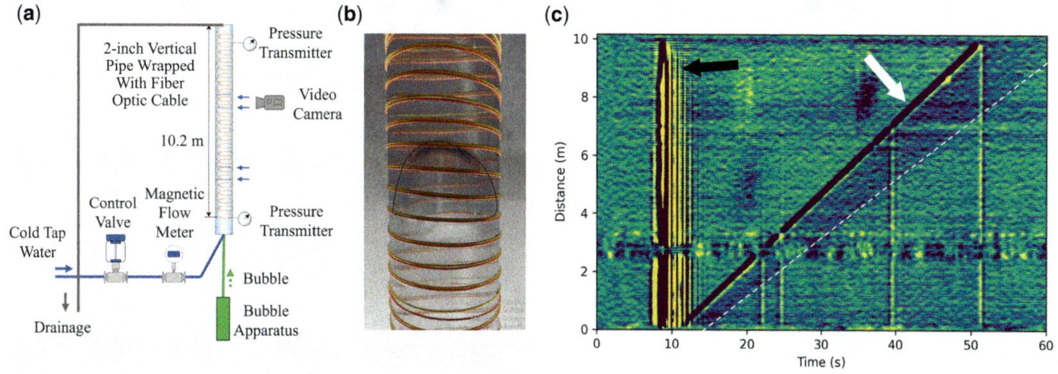

Figure 1.26. (a) Design of vertical flow loop facility with attached bubble apparatus, (b) example Taylor bubble inside pipe, and (c) processed DAS record. The vertical axis is the distance along the pipe. The horizontal axis is time in seconds. The white arrow indicates the track of the rising bubble which is the dark colored event propagating from approximately 10 s at the bottom up to 50 s at the top. The black arrow indicates tube waves. (Modified from Figures 1a, 2b, and 3c of Titov et al., 2022.)

Titov et al. (2020, 2022) present DAS data acquired while flowing fluid and gas bubbles in a pipe using a laboratory flow loop. Their goal is to be able to quantify the acoustic noise signals in the pipe with the properties of the multiphase fluid. Figure 1.26a shows the apparatus used to create Taylor 2 bubbles in a flowing fluid stream inside a pipe. Wrapped around the outside of the pipe is an orange fiber optic cable which was then attached to a DAS interrogator unit and recording system. Figure 1.26b shows a photograph of one of the bubbles inside the transparent pipe, where the orange fiber optic cable can be seen on the outside of the pipe. Figure 1.26c shows the processed DAS record revealing tube waves as vertical lines and the track of the bubbles as a diagonal line. The authors have studied the propagation speeds with respect to the bubble size and number of bubbles.

1.5. Summary

I've shared a few of the many applications for fiber optic DAS technology in this chapter. I hope that at least some are both new to you and inspiring. Earthquake detection using existing fiber optic cable systems, including land and subsea telecommunication systems, are providing an opportunistic, cost-effective method of detecting events and characterizing the structure of the earth, as well as warning about pending tsunamis. Monitoring volcanos with DAS technology has been demonstrated for the Hunga Tonga volcano (FOSA, 2022). The low-cost ability to install DAS systems for monitoring facilities such as buildings, roads, and pipelines make it ideal for safe-guarding important infrastructure from intrusion by vehicles, people, and animals.

The use of DAS technology for seismic exploration and carbon sequestration monitoring is growing rapidly. Surface seismic methods can use fiber that is simply laid out on the ground, which is very inexpensive but may suffer from poor acoustic coupling. Solutions such as burying the fiber can greatly improve the coupling, but that may be somewhat hampered by permitting and environmental issues. Using DAS technology in borehole wells is much easier because it is placed inside the well and no additional disruption of the surface is required. Several different methods exist to install the fiber optic cable in the well either for permanent or temporary monitoring purposes.

Fiber optic cables installed in a well can be used to acquire seismic data from active controlled seismic sources (e.g., air guns and vibrators) or can be used to monitor passive sources of sound. The VSP data sets are acquired using active sources to be able obtain the detailed structure of the subsurface. Microseismic data can be recorded by DAS in vertical and horizontal wells to help locate and characterize the extent of hydraulic fracturing or subsidence. Low-frequency strain rate data acquired in a monitor well are being used to map the location and extent of individual fracture groups during hydraulic fracturing. Flow noise from each perforation can indicate the quality of that perforation in a well by monitoring with a fiber optic cable using DAS. Finally, I have presented an example of DAS being used to characterize the progression of bubbles in flowing liquid inside a pipe.

In this first chapter, I've shown many examples of DAS technology being used for amazing applications – many of which I'm sure are likely new to you. Because the technology can be used for so many diverse purposes, it an important topic to learn more about for your professional career even if you aren't currently utilizing it. In the following chapters, I will describe the technology and how it can be used effectively.

Chapter 2 Overview of distributed acoustic sensing technology

I ask myself, wonderstruck, what prodigious thing will he do today, this Salvador Dali.

– Salvador Dali

2. Laying the foundation for understanding distributed acoustic sensing

In the first chapter, I introduce many of the diverse applications of DAS technology without explaining how it is different from conventional seismic technology or describing how it works. In this chapter, I'll present a high-level overview of DAS and then in subsequent chapters delve more systematically into some of the details. Let's dig in!

2.1. *Comparing traditional and fiber optic technology*

We are at a crossroads where many of the applications I've shown using physical sensors, including hydrophones, geophones, and seismometers which have been developed and deployed for more than 100 years, are being augmented or completely replaced with fiber optic cables using DAS technology. At the moment, we are living in a convergent environment with individual electrical sensors being used in some applications and fiber distributed acoustic sensing in others. The move toward replacing discrete sensors with continuous fiber is appealing for several reasons I detail next.

Figure 2.1. Güralp 40T Broadband three-component earthquake seismometer. (Courtesy Güralp.)

2.1.1. Cost factors

The first compelling reason to move to DAS technology is the cost factor – spending the appropriate amount of capital to solve the required problem in the most effective manner.

Conventional earthquake seismometer instruments (Figures 2.1 and 2.2) are highly calibrated and expensive (approximately US $10,000 each). It is cost effective to have a few highly sensitive seismometers distributed across a wide area to be able to characterize global earth features.

Figure 2.2. Interior view of a vintage Willmore vertical component seismometer. (Courtesy Seth and David Willis.)

In contrast, we can characterize features of the shallow (<12 km) subsurface to identify water, hydrocarbon, geothermal, and carbon sequestration reservoirs by using active seismic sources and relatively inexpensive (approximately US \$100–\$1000 each) geophones (Figure 2.3) placed either on the surface or in boreholes.

As of this writing, fiber optic cable is on the order of US \$0.70 per meter. From the raw materials point of view, you can instrument about 150 meters of dense (about every 1 meter) fiber optic sensor coverage for the price of one geophone. For surface seismic exploration applications, the geophone spacing is typically one every 25 meters. So, for 150 meters of sensor coverage, it would cost about \$600 for the six geophones and only \$100 for the corresponding fiber optic cable. This is a factor of six in savings with an increase in sensor density of 25 for the fiber optic cable over the geophone solution.

2.1.2. Sensor density

The second compelling reason to move to DAS technology is sensor density. With fiber optic cables, we can make a measurement essentially every 1 meter. (In some cases, this can be decreased to sub-meter spacing.) For waves which travel at slow velocities, the increase in sensor density allows for unaliased, or in other words fully sampled, recording of the waves. This allows a more accurate characterization and delineation of the different wave types.

Figure 2.3. Cutaway view of a one-component surface seismic geophone.

2.1.3. Temperature sensitivity

The third compelling reason is the better tolerance to high temperatures. Fiber optic cables can typically withstand much higher temperatures (–50°C to 400°C) and pressures. In contrast, geophones typically have maximum temperature ratings of –40°C to 200°C. For surface applications, typical atmospheric ambient temperatures range from –30°C to 49°C, which allows for either geophones or fiber optics to be used. However, for borehole environments, subsurface temperatures can exceed these ranges. The thermal gradient of the upper 100 km of the crust is typically between 15°C/km and 30°C/km (Earle, 2019). For a deep well (9 km), plausible temperatures exceeding 280°C would easily render geophones inoperable while the fiber optic solution is still well within its operating temperature range.

2.1.4. Electrical and mechanical reliability

The fourth compelling reason is improved electrical/mechanical reliability. Fiber optic cables have no moving parts to wear out. In contrast, geophones have moving parts that are susceptible to damage. Fiber optic cables have two intrinsic weaknesses which can be managed. Fiber optic cables must not be bent past a critical radius, otherwise the light will no longer be fully contained in the glass and will leak out. Thus, procedures need to be implemented for handling the fiber optic cables in the field to ensure they are given the proper bending radius. They also can be damaged by contact with hydrogen, so that must be managed (Maida, 2008).

2.1.5. Ease of deployment

The fifth compelling reason is the ease of deployment. The diameter of an optical fiber is typically 125 μm, which is about the width of a strand of hair, and thus its physical dimension is very small. Of course, it must be housed inside of a set of protective coatings and tubes, which makes it a little larger in diameter but still much smaller and less bulky than a geophone. Thus, the fiber optic cable can be easily deployed in places where geophones would be much more difficult and costly.

2.2. Overview of the DAS technology

The DAS technology is composed of five key aspects:

1. The fiber
2. The fiber deployment method
3. The interrogator
4. The gauge
5. The conversion of the native phase measurement to strain or strain rate.

I'll give a quick overview of the technology in the following sections and then cover them in more detail later.

2.2.1. The fiber optic cable

A fiber optic glass filament consists of a glass core surrounded by cladding, as shown in Figure 2.4. Light travels in the glass core and the surrounding cladding provides the optical mechanism to keep the light trapped inside the core. Using ray theory from geometrical optics, the cladding will keep light inside the fiber for all ray angles smaller than the critical angle of refraction, θ_c, given by (equation 1.1 of Thévenaz, 2011)

$$\theta_c = \cos^{-1}(n_{clad} / n_{core}), \tag{2.1}$$

where n_{clad} and n_{core} are the indices of refraction of the cladding and core, respectively, and θ is the angle a ray makes with respect to the cladding. An angle of 0 indicates the

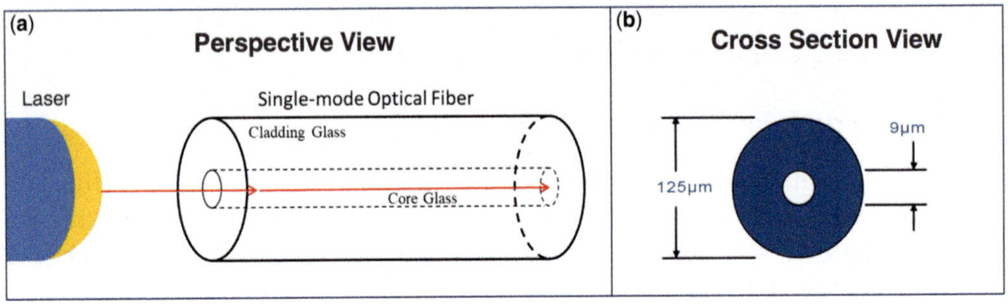

Figure 2.4. Optical fiber construction for a single mode fiber. (Modified from Willis et al., 2018.)

ray is propagating in the axial direction of the fiber. The cladding also provides additional mechanical strength. A plastic jacket is placed around the outside of the cladding to absorb stray light and to protect it from environmental interactions that might degrade the fiber.

Fiber optic cables typically contain several optical fibers which together are encased in metal and/or plastic tubes which protect them from mechanical damage. A gel is typically inserted around the fibers which separates them from the tubing and allows them to slide independently of each other. Figure 2.5 shows a diagram of the cross section of a fiber optic cable with a fiber in the center surrounded by gel. Around this is a stainless-steel inner tube and an aluminum inner tube, which then are surrounded by an outer tube and a protective encapsulation. An example FIMT (fiber in metal tube) is shown in Figure 2.6, wherein several optical fibers are enclosed in the protective metal tubing.

Each fiber in the fiber optic cable can be used to measure the motion of the ground. The DAS measurement is almost exclusively sensitive to the longitudinal stretching or compressing of the fiber. This makes it a one-component measurement aligned with the direction of the fiber itself. We typically refer to the length of fiber that we want to use for measurements as the "sensing fiber," as shown in Figure 2.7. The fiber cable that is being used to obtain the light from our measuring instrumentation to the sensing fiber is typically called the umbilical or surface cable. While we may make measurements in those sections of the fiber optic cable, the portion of the cable to be used for recording the desired signals is called the sensing fiber.

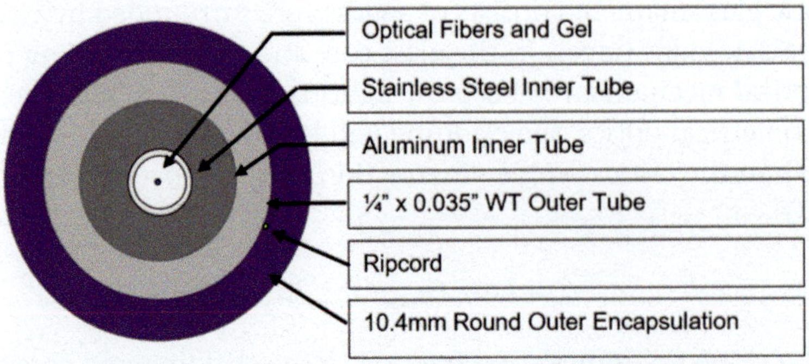

Figure 2.5. Schematic of a fiber optic cable. (Modified from Ellmauthaler et al., 2020b.)

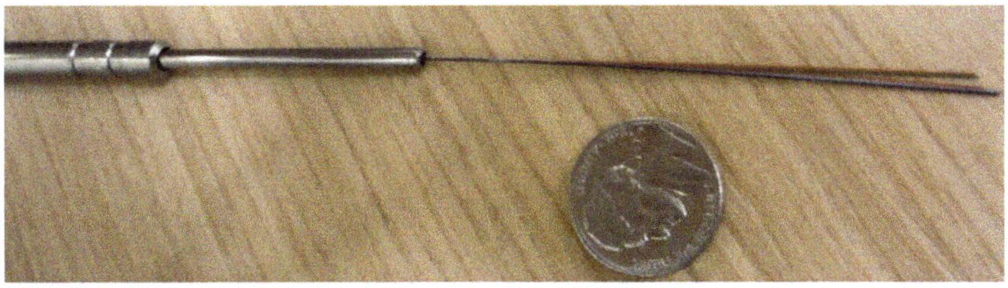

Figure 2.6. Example of a FIMT showing the four fibers inside the metal tubing. A United States nickel is shown for a size reference. (Courtesy Halliburton.)

2.2.2. Fiber deployment methods

Two different styles of fiber optic cable deployment are used. The cable can be positioned on or near the surface. Or alternatively, the cable can be lowered into a borehole.

When the cable is deployed on the surface, it is difficult to achieve good coupling because the cable is not very heavy. Figure 2.8a shows a fiber optic cable laying on a grassy field where it is held off the ground by grass and twigs. One fairly costly approach is to lay something on the cable, such as the wood shown in Figure 2.8b, in an attempt to increase the coupling quality. However, it is typically buried in shallow trenches (Figure 2.9). Studies also have noted that good coupling of the cables can be achieved if they are pushed into the soil (Martin, 2021), laid on the ground surface (Spikes et al., 2019), or in cold locations laid on the ground with water poured over so that it freezes in place (Yang, 2021). The deployment of surface fiber is relatively involved and cumbersome when there are infrastructure obstacles such as roads, buildings, and near surface rocks. However, these issues can be mitigated if there are existing fiber optic cables deployed in telecommunication tunnels in the desired location that can be used (Martin et al., 2017).

For many exploration seismic applications, the fiber optic cable is deployed in a well using one of several methods, as illustrated in Figure 2.10:

1. The cable can be strapped to the outside of casing and cemented into the well (Figures 2.10a and 2.11).
2. It can be strapped to the outside of production tubing (Figure 2.10b).
3. It can be deployed as a component of a wireline cable (Figure 2.10c).

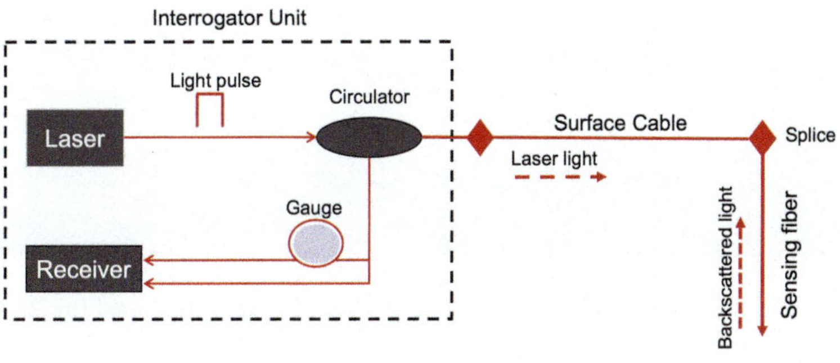

Figure 2.7. Diagram of the interrogator unit, surface cable, and sensing cable.

Figure 2.8. (a) Fiber optic cable laid on the ground, and (b) wood laid over the fiber optic cable to increase coupling.

4. The fiber can be pumped into a well, which is especially useful in shallow SAGD (steam-assisted gravity drainage) wells (Figure 2.10d). A very inexpensive option is to deploy the glass fiber directly into the well by gravity in a temporary disposable manner (Higginson et al., 2017; Green, 2021; Figures 2.10d and 2.12).
5. Another way to deploy fiber into a well is via using coiled tubing (Ellmauthaler et al., 2019). The sensing fiber optic cable is threaded through the coiled tubing and deployed into the borehole (Figures 2.10e and 2.13).

Using whichever method chosen to deploy it, the fiber itself is the basis for sensing the motion of the earth.

Figure 2.9. Example of trenching in a fiber optic cable: (a and b) trenches and (c) installing the fiber. (Modified from Figure 3 of Ajo-Franklin et al., 2017.)

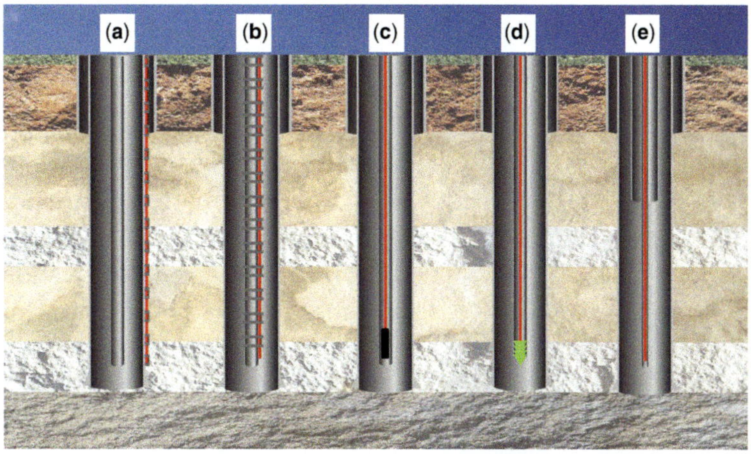

Figure 2.10. Fiber (red) deployment methods: (a) outside of casing, (b) on tubing, (c) via wireline, (d) pumped down or free hanging, and (e) in coiled tubing.

2.2.3. The interrogator unit

The interrogator unit (IU) is probably the most complicated piece of equipment in the DAS system. As illustrated in Figure 2.7, the deployed fiber optic sensing cable is connected to the surface cable, which in turn is connected to the IU within the recording truck or doghouse. The IU contains a laser which is the source of light that is pulsed into the sensing fiber. The laser light then travels down the sensing fiber. In all fiber optic glass, there are small impurities or changes in the refractive index of the glass. These changes in the refractive index will cause the down-going laser light to backscatter back up the fiber toward the IU. Because these changes in refractive index occur nearly continuously over the entire length of the fiber, the backscattered light will carry with it information about the acoustic vibrations that the fiber is detecting from the entire length of the deployed fiber.

2.2.4. The gauge

The gauge is a device used in the interrogator to implement optical interferometry. While there are many different ways to implement this optical interferometry (Hartog, 2017), I will describe one that seems the most straightforward. The laser in the IU (as shown in Figure 2.7) emits a coherent pulse of light composed of a single frequency which shines down the sensing fiber. Tiny changes in the fiber optic glass, due both to imperfections and miniscule changes in the index of refraction, cause the laser light to be backscattered up the fiber and into the IU again. Light from two different places in the sensing fiber are brought together using an optical time-delaying strategy. The backscattered light coming from the sensing fiber is split into two paths within the IU – the direct path and a time-delayed path.

The time delay is created by a loop of fiber of a fixed length called the gauge length, as shown by the purple disk in Figure 2.7. Then, the light from these two paths are merged back together. Because the light will have traveled two different path lengths, the light will likely be out of phase from each other. They will either constructively or destructively interfere.

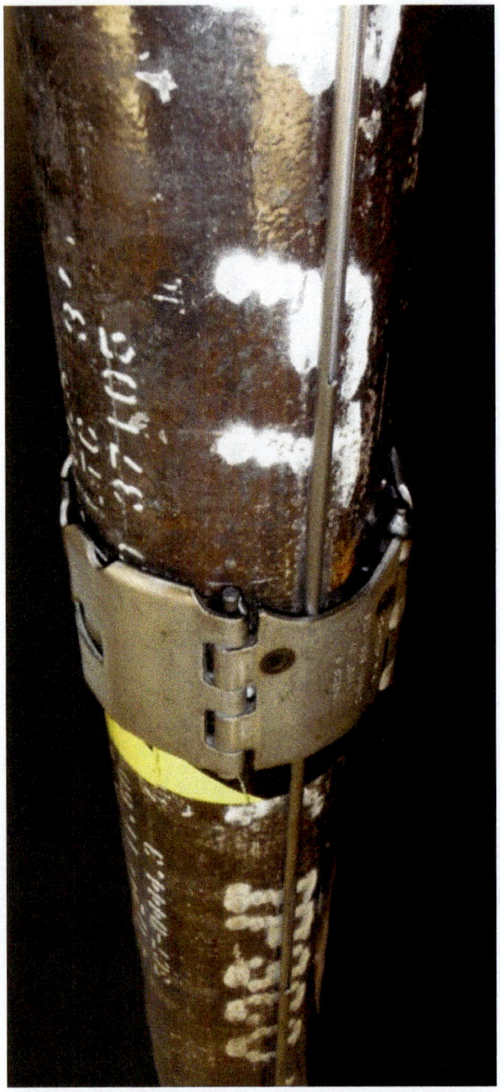

Figure 2.11. Fiber optic cable strapped outside of casing. (Courtesy Halliburton.)

This interferometric process is most easily demonstrated using microscope slides and a light source. Figure 2.14a shows a diagram of two microscope slides placed on top of each other. Two reflected light rays are shown – one from the bottom of the top slide and the other from the top of the bottom slide. The reflection from the bottom slide will have a 90° phase change relative to the reflection from the top slide. In addition, the distance that the two rays will travel are different by an amount equal to the distance between the slides. Thus, when the two rays are seen, either by our eyes or a camera, for example, they will interfere, forming fringes or rings which are dark when out of phase, and bright when in phase.

I set up this demonstration in Figure 2.14b. The top image shows the interference patterns when the slides are simply resting against each other. There are more or less vertical patterns of lighter and darker zones. In the bottom image, I have pressed the slides together with my thumb on the left side. Now, we see that the patterns change over the entire slide, indicating a bending of the slides relative to each other. The distance between one complete cycle (white to white fringe, or black to black fringe) is directly related to the difference in the distance between the slides at those two points. One cycle indicates the slides are now one half period of light farther apart, because the light travels down in the air gap to the bottom slide and back up to the upper slide. One half period is equal to approximately 28 μm (2.8×10^{-5} m = (speed of light)/(mean frequency of light)/2 = (3×10^{8} m/s)/(536×10^{12} Hz)/2). Thus, we can accurately determine the relative distance the slides are from each other.

Returning to our DAS interrogator, we can measure the phase angle of the light between the straight path of the backscattered light and the backscattered light delayed through the gauge. Using optical interferometry, we can very accurately determine the relative difference in distance they traveled from the phase angle of the light that is measured. Because the backscattered light from each meter of the sensing fiber returns to the IU at a different time, a measurement of the phase angle difference for each meter of the well can be determined. When a seismic signal hits the deployed sensing fiber, it strains the fiber locally to be either very slightly longer or shorter. Thus, the

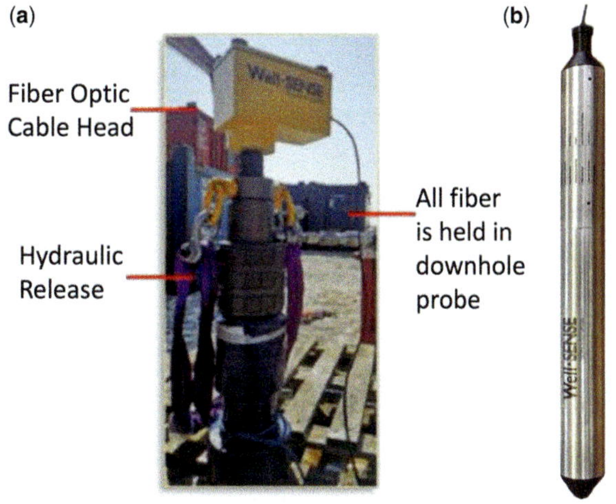

Figure 2.12. (a) Disposable fiber deployment system, and (b) the "torpedo" containing the fiber which is dropped into the well. (Modified from Figures 1 and 2 of Green, 2021.)

distance the light has traveled from the backscattering sites will change. This resulting change in distance will change the measured phase angle and thus provide a way to detect this seismic signal.

2.2.5. Strain and strain rate

After measuring the phase angle difference of the light using the interferometer, it next must be converted to the seismic response. The phase angle difference that is measured for each meter along the sensing fiber can be converted to a seismic signal using some simple math. This conversion process has been standardized by the SEAFOM organization (e.g., SEAFOM, 2018). However, the main notion is that the phase angle is converted to a measure of the deformation of the fiber. The unit of strain (or strain rate) of the fiber is created from the measured phase difference.

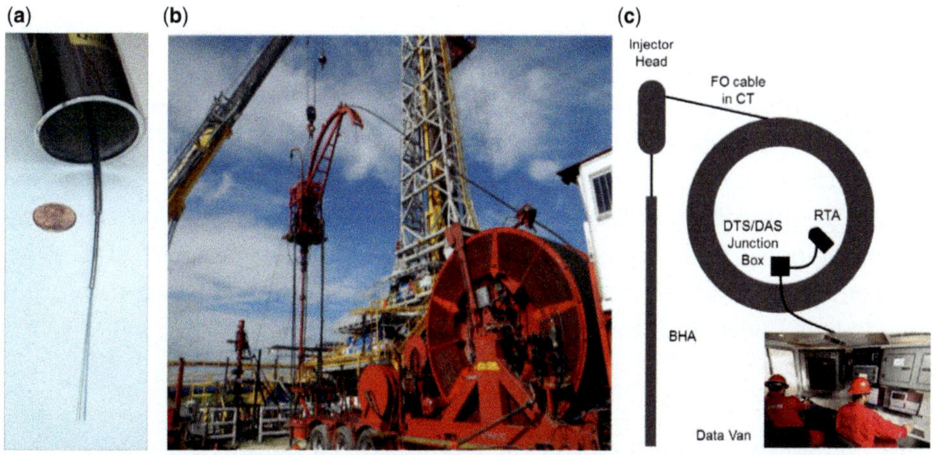

Figure 2.13. (a) Fiber optic cable inside of coiled tubing, (b) the coiled tubing deployment system, and (c) diagram of the components of the coiled tubing system. (Courtesy of Halliburton.)

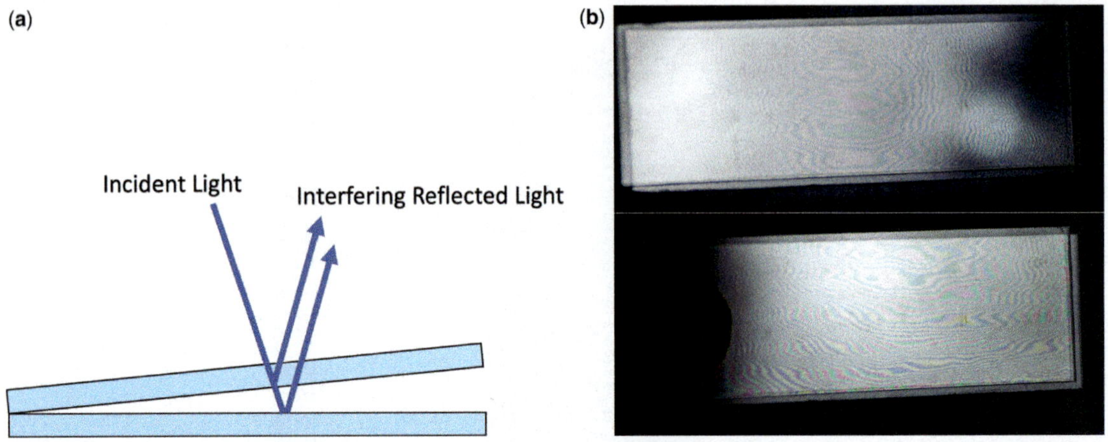

Figure 2.14. (a) Edge view diagram of two microscope slides being illuminated by an incident light. The reflected light from the top and bottom slides interfere to form fringes of lighter and darker light. (b) Resting state of two slides showing light and dark light fringes (top), and after applying pressure on the left edge of the slides, causing the pattern change that reflects the new distance between the two slides (bottom).

The relationship between the measured phase angle for the two light paths in the interferometer is given by (SEAFOM, 2018)

$$\epsilon = \frac{\lambda d\phi}{4\pi n GL \xi}, \tag{2.2}$$

where ϵ = strain, λ = operational optical (vacuum) wavelength of the DAS system, n = the refractive index of the sensing fiber, GL = the gauge length, ξ = the photo-elastic scaling factor for longitudinal strain in isotropic material ($\approx$0.78), and $d\phi$ = the phase angle difference of the light measured in radians.

The strain rate is simply the time derivative of strain given by

$$\dot{\epsilon} = \frac{d\epsilon}{dt}. \tag{2.3}$$

All of the terms on the right side of equation 2.2 are constants except for $d\phi$, the phase difference. This means that the strain rate is controlled by the change over time of the phase difference of the light from the two paths.

Typically, geophysicists are accustomed to measuring seismic waves in terms of displacement, velocity, and acceleration. They may be unfamiliar with strain (or strain rate) measurements. However, strain (and strain rate) can be converted to more familiar units if needed, as I will describe in the next chapter.

2.3. Summary

In this chapter, I've described the key elements that comprise a fiber optic DAS system: 1) the sensing fiber, 2) the fiber deployment method, 3) the interrogator unit, 4) the gauge, and 5) the conversion of the measurement to strain (rate). The sensing fiber has a center core which traps the light to make it remain within the glass. Then, the optical fiber is packaged into a multi-layered cable system to protect it and make it easy to transport and install. The fiber cable can be deployed either on or near the surface of the ground or it can be lowered into a well bore. From these different vantage points, DAS data can be acquired using an interrogator unit. The heart of the interrogator is the gauge which allows optical interferometry to be performed. Optical interferometry is a method which allows the acoustic energy hitting all along the sensing fiber to be translated into a measurement of strain (rate). The measured strain (rate) forms the basis for using the recorded data as a finely sampled seismic wavefield.

Chapter 3 Definitions of strain and strain rate

Sometimes I feel like tap-dancing, screeching, unscrewing light bulbs, pulling curtains, combing hair, doing knee bends, handstands and turning somersaults out there.

– *Ilie Nastase*

3. Physics of the measurements

In order to understand the difference between what is used by most geophysicists and what DAS is measuring, let us first very briefly review the physics relating the measurements.

3.1. Strain and strain rate defined

Strain is defined "as the change in position of a point with respect to its neighboring points divided by the distance between its original point and the reference points" (Officer, 1958, page 3). As such, it can be thought of the decimal percentage change in the dilation or contraction of an infinitely small volume around that point. In three dimensions, strain can be characterized by a tensor consisting of nine components which are defined on the six faces of a cube of material (Fung, 1977, page 126). The DAS IU is only able to measure one component of strain along the axial (length-wise) dimension of the fiber. We will define that direction to be along the z-axis of our coordinate system. (For fiber in a vertical well, the z-axis would correspond to the depth axis. For fiber on the surface, the z-axis corresponds to the distance along the ground.) Thus, the strain component that the DAS IU measures is ε_{zz}, or in other words the strain from an applied force in the z-direction measured in the z-direction. Note that, if the reference to z is omitted in the equations, we will assume it to be in the z-direction because the fiber can't measure the strain in any significant way along any other direction.

Figure 3.1 shows a bar at two states of time – the beginning (top) and at a later time (bottom) with an altered length after stress has been applied to the axial, z, direction. If the length of a bar is L_0, and the size of the bar in that direction after being stretched by a force is L, then the strain ϵ is given by (Fung, 1977, eq. 5.1-1)

$$\epsilon = \frac{L - L_0}{L_0}. \tag{3.1a}$$

In this illustration, the bar is longer after the stress has been applied, which means that $\epsilon > 0$ and the applied stress was in tension. The bar has been pulled, not compressed, and so it is longer. Strain can be thought of as the dimensionless, decimal

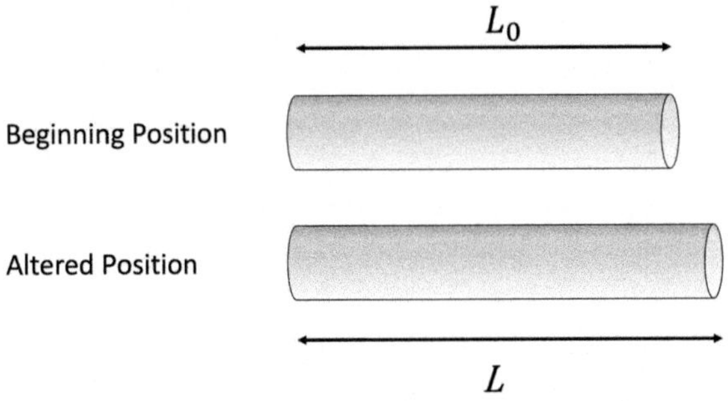

Figure 3.1. Diagram to help define strain. The top image is a bar with length L_0 at the beginning. The bottom image is the bar which has been altered by an applied stress and is now length L.

percentage change in the length. The infinitesimal strain value at a single point is defined as if the length L_0 in equation 3.1a is shrunk to zero. The practical difficulty with measuring strain is determining the original "resting" position, L_0, of the bar.

The change of strain with time is the strain rate, $\dot{\epsilon}$, which is simply the time derivative the strain, ϵ, given by

$$\dot{\epsilon} = \frac{d\epsilon}{dt}. \tag{3.1b}$$

3.2. Strain and strain rate approximated using DAS

The interferometer in the IU detects minuscule changes in the tension or compression all along the length of the deployed fiber. These changes typically are sampled and recorded for every meter along the fiber. To do this, the IU uses a time-delay mechanism to combine the backscattered light from portions of the fiber that are spaced apart by the gauge. The DAS strain response, ϵ^{DAS}, therefore measures the average strain over the distance of the gauge length, GL. For example, a fiber is 5000 m long and deployed in a well or on the surface. This means that the IU will create 5000 samples of strain (or strain rate) all along the fiber for every pulse of the laser. Each of these samples will be an estimate of the average of strain (or strain rate) over the gauge length.

To be more precise, for seismic waves incident on the fiber, the average measured strain, created by the integration over the gauge length, is given by (modified from Bakku, 2015, eq. 3.5)

$$\epsilon^{DAS} = \int_{z-\frac{GL}{2}}^{z+\frac{GL}{2}} \frac{\epsilon}{GL} dz' = \epsilon \, sinc\left(\frac{k_z GL}{2}\right). \tag{3.2}$$

Thus, the measured DAS strain, ϵ^{DAS}, is not the actual strain, ϵ, in the fiber, but is instead the strain multiplied by an extra sinc function term, which is dependent upon the angular wavenumber, k_z, in the z- (axial fiber) direction and the gauge length. Much more will be discussed about this in Chapter 6.

The DAS strain rate, $\dot{\epsilon}^{DAS}$, is the simple time derivative of the strain, given by

$$\dot{\epsilon}^{DAS} = \frac{d\epsilon^{DAS}}{dt}. \tag{3.3}$$

While appearing straightforward in these two equations, it is not, and more details of these computations will be given in Chapter 6.

3.3. Simulating strain and strain rate data from velocity data

Considering only the axial direction of the fiber, the displacement of a point along the fiber can be described by the function w, oriented along the z-axis. For a stress applied axially to the fiber, the resulting strain can be found from (Officer, 1958, eq. 1–3)

$$\epsilon_{zz} = \frac{\partial w}{\partial z}. \tag{3.4}$$

Equation 3.4 says the strain is simply the spatial partial derivative of the displacement, w, in the z-direction, along the axial direction of the fiber.

From first principles, the particle velocity in the z-direction, V_z, is the time derivative of the displacement w. Thus, we can write

$$V_z = \frac{dw}{dt}. \tag{3.5}$$

For completeness, the acceleration A_z is a further time derivative:

$$A_z = \frac{dV_z}{dt} = \frac{d^2w}{dt^2}. \tag{3.6}$$

For now, let's ignore the geophone's instrument response due to the resonance of the spring suspending the mass. To convert from a particle velocity measurement to strain, we combine equations 3.4 and 3.5 to form

$$\epsilon_{zz} = \frac{\partial}{\partial z}\left(\int V_z dt\right). \tag{3.7}$$

Strain rate, $\dot{\epsilon}$, is simply the time derivative of strain, so

$$\dot{\epsilon} = \frac{d\epsilon}{dt}. \tag{3.8}$$

This means that the conversion from particle velocity, V_z, or acceleration, A_z, to strain rate is simply

$$\dot{\epsilon}_{ZZ} = \frac{\partial V_z}{\partial z}. \tag{3.9}$$

$$\dot{\epsilon}_{ZZ} = \frac{\partial}{\partial z}\int A_z dt. \tag{3.10}$$

Therefore, it is easy to simulate strain rate data from particle velocity data by simply taking a spatial difference of the particle velocity data via equation 3.9. The spatial differencing should be performed using the gauge length. Taking the differences evenly spaced around each location by using centered differences, this becomes

$$\dot{\epsilon}_{ZZ}^{DAS} = \frac{V_z(z+GL/2)-V_z(z-GL/2)}{GL}. \tag{3.11}$$

Taking the difference over the gauge length distance inserts the sinc function back into the response of the DAS strain rate from equation 3.2. This relationship has been shown by multiple authors (Mateeva et al., 2014; Bakku, 2015; Daley et al., 2016; LeBlanc et al., 2018). You can practice the conversion from particle velocity to strain rate using a MATLAB gui (Willis, 2022b).

An additional aspect needs to be considered when comparing geophone and strain (rate) measurements, and that is the geophone instrument response.

3.4. Simulating strain and strain rate data from geophone data

A geophone attempts to measure the particle velocity. However, because of its construction as a physical device, it has resonances that add complexity to its response. Hons et al. (2007) and Hons (2008, eq. 1.16) show that the impulse response of a velocity geophone, $R(f)$, is given by

$$R(f) = -S_G \frac{f^2}{-f^2 + 2j\lambda f f_0 + f_0^2}, \tag{3.12}$$

where f is frequency, S_G is the sensitivity constant of the geophone, λ is the critical damping of the geophone, and f_0 is the center frequency of the resonance of the spring mass system.

Figure 3.2 shows the normalized velocity response of a geophone with $f_0 = 10$ Hz and $\lambda = 0.7$. For these parameters, the response of the geophone is relatively flat above 3 Hz. However, below 3 Hz the response drops off relatively quickly. (Of course, the corner roll-off frequency is dependent upon the value of f_0.)

What this means is that, if we try to simulate DAS data from equation 3.11, assuming the geophone is measuring velocity, at low frequencies we will be

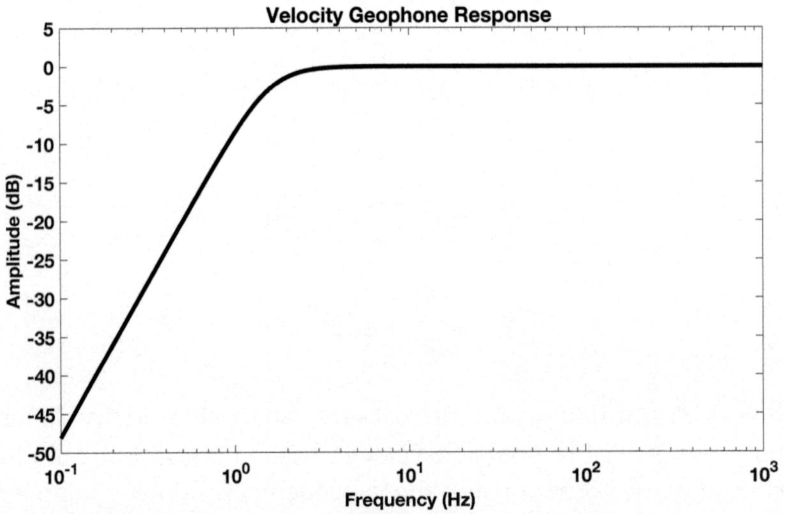

Figure 3.2. The velocity response of a geophone with $f_0 = 10$ Hz and $\lambda = 0.7$.

underestimating the DAS signal level. However, at higher frequencies, equation 3.11 will be a good approximation. For a complete simulation of the strain and strain rate data from geophone data, the geophone instrument response should be removed by deconvolution. This is conceptually given by

$$\Delta_s = \mathcal{F}\left\{\frac{D_z(z+GL/2) - D_z(z-GL/2)}{GL}\right\}, \tag{3.13a}$$

$$\epsilon^{DAS} \approx \mathcal{F}^{-1}\left\{\frac{1}{R(f)}\Delta_s\right\}, \tag{3.13b}$$

where $\mathcal{F}$ indicates the Fourier transform of the variable to the frequency domain, Δ_S is the Fourier transform of the differences of D_z, and D_z is the integrated over time response of the geophone, G_z, given by

$$D_z = \int G_z dt. \tag{3.14}$$

In a similar manner, we can obtain the strain rate directly from the geophone response from

$$\Delta_{sr} = \mathcal{F}\left\{\frac{G_z(z+GL/2) - G_z(z-GL/2)}{GL}\right\}, \tag{3.15a}$$

$$\dot{\epsilon}^{DAS} \approx \mathcal{F}^{-1}\left\{\frac{1}{R(f)}\Delta_{sr}\right\}, \tag{3.15b}$$

where Δ_{sr} is the Fourier transform of the differences of G_z.

This means that we can compare DAS strain and strain rate data to geophone data by converting the geophone data to strain and strain rate using equations 3.13 and 3.15, respectively. For the example geophone characteristics in Figure 3.2, removing the geophone response would have little effect on the simulated DAS signal above 3 Hz. This is straightforward for raw seismic records, such as shot gathers (or common receiver gathers), because there is a direct way to perform the spatial differencing required of the geophone data. However, after geophysical products are created, such as angle gathers, corridor stacks, CDP transform images, or migrated images, it is not as clear how this operation should be performed.

3.5. Simulating particle velocity data from strain (rate) data – the temporal integration method

Exploration and earthquake seismology nearly always use seismometers and geophones to measure the motion of the earth. Thus, the natural unit of motion has become particle velocity (and sometimes displacement or acceleration). To compare conventional seismic velocity measurements with DAS strain (rate) data, it is helpful to be able to convert strain data to particle velocity units. Equation 3.11 provides a very simple differencing scheme allowing us to convert velocity measurements to strain rate data. Let's try to reverse that now, converting strain (rate) data to the particle velocity.

Daley et al. (2016, eqs. 4 and 5) show that, for a single seismic wave, traveling at an apparent, single constant velocity, c, the transformation to velocity from strain is simply

$$V_z = \pm c\epsilon. \tag{3.16}$$

The sign is selected appropriate for the direction of the propagating wave, i.e., positive for forward propagation and negative for backward propagation. As pointed out by the authors, this approach will not work for a superposition of waves propagating at different velocities and/or if they exhibit dispersion. However, the beauty of this approach is that each arrival in the DAS strain data domain can be multiplied by a simple scalar representing the local velocity of the wave.

The corresponding formula for the strain rate is

$$V_z = \pm c \int \dot{\epsilon} dt. \tag{3.17}$$

We will call the method using this equation the temporal integration method. Let's see how well this works on a synthetic data example. Figure 3.3 shows synthetic vertical component velocity, V_z, traces in the first column, the corresponding DAS strain rate traces in the second column, and a comparison of the DAS strain rate traces converted to V_z (in red) with the V_z traces (in black) in the third column. The synthetic V_z trace uses a Ricker source wavelet with a center frequency of 70 Hz, a formation velocity of 2500 m/s, and a source to receiver distance of 2500 m, conveniently making it arrive at 1 second. The DAS strain rate trace is created using equation 3.11.

The V_z traces in the first column (Figure 3.3a, 3.3d, and 3.3g) are identical. Equation 3.9 defines an exact relationship on how to go from velocity, V_z, to strain rate. However, in practice, we cannot use this equation directly because we can't shrink the depth increment to "zero" to take the derivative. Thus, we must use equation 3.11. This means that, for appropriately small gauge lengths and fast apparent velocities, the strain rate computed using equation 3.11 will be valid. However, for inappropriately large gauge lengths or slow apparent velocities, the relationship will have artifacts.

The strain rate trace for a 5 m gauge is shown in Figure 3.3b, which looks like a time derivative of the V_z trace in Figure 3.3a and not like a Ricker wavelet. Making the gauge length larger to 15 m (in Figure 3.3e) creates a somewhat stretched out version of the 5 m case. Further increasing the gauge to 40 m (Figure 3.3h) seems to reveal the negative V_z wavelet centered at 0.992 s, and the positive V_z wavelet centered at 1.008 s in the DAS strain rate trace.

We next apply equation 3.17 to the DAS strain rate traces in the center column to create estimates of the V_z trace. For the 5 m gauge length traces in the first row (see Figure 3.3c), we have an excellent match between the V_z trace (black) and the converted DAS strain rate trace (red). Applying this same approach to the 15 m gauge trace in the middle row gives the converted trace (red) in Figure 3.3f. The conversion has done a reasonable job changing the derivative-like Ricker wavelet in Figure 3.3e into something similar to the Ricker wavelet V_z. However, we see that the amplitude has been diminished and the wavelet slightly broadened.

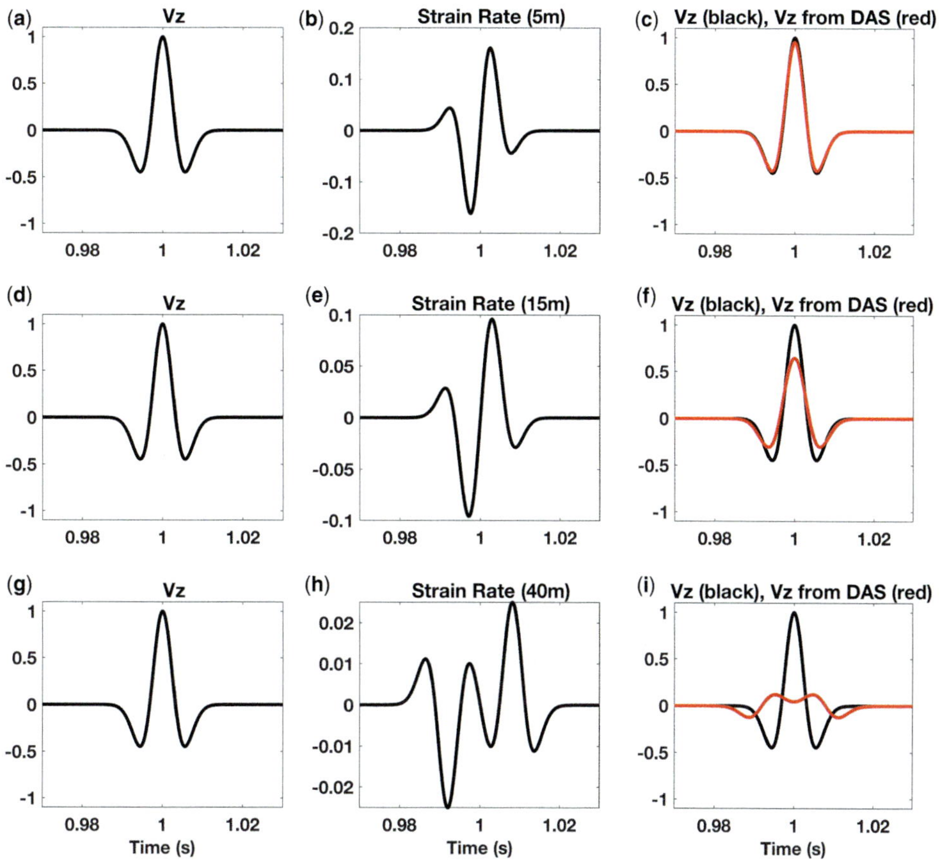

Figure 3.3. Conversion from strain rate to V_z using the temporal integration method. Top row: gauge length 5 m. Middle row: gauge length 15 m. Bottom row: gauge length 40 m.

Finally, let's look at the result of converting the 40 m gauge length trace to V_z in Figure 3.3i. The resulting wavelet is much too weak in amplitude and the shape is severely distorted. As we will discuss later, the effect of the gauge is pronounced and detrimental for slower apparent velocity events and high frequencies. The bigger the gauge, the more distortion.

You can practice converting strain rate to particle velocity using a MATLAB gui (Willis, 2022c).

Now that I've shown an empirical example of the effect of applying equation 3.17 to DAS strain rate data, let's examine that equation in more detail. The sign in front is a "user" selectable option which allows us to flip the polarity of the up-going waves to match the polarities of particle velocity and strain rate data. Next the velocity, c, term is a simple scalar which boosts the amplitude of the strain rate trace back to the appropriate level for V_z data. The integration over time attempts to remove the spatial derivative operation that was performed by the gauge. This integration is the only part of this formula that changes the wavelet shape. From this example, we see that matching the gauge size for the expected apparent velocity and frequency content of the data will result in an effective conversion to velocity from strain rate. However, this formula will not be effective if the gauge size is not selected appropriately.

What happens to events traveling at a different apparent velocity than the one selected to be converted? On a typical DAS record, there will many different waves propagating. On a VSP record, there likely will be P-waves, S-waves, and tube waves. On a surface seismic DAS record, there will be surface waves and body waves. So, if equation 3.17 is used to convert the DAS strain rate data to vertical particle velocity (V_z), then a single apparent velocity must be used for each event. As mentioned, the sign selected will simply flip the polarity of the event. The temporal integration will change the wavelet shape. But, the velocity used for the correction will impart an erroneous amplitude scaling for an event traveling at a velocity different from the one selected for conversion. Thus, the "other event" will either have an enhanced or diminished amplitude than it should. This shortcoming is addressed by the spatial integration method described in the next section.

3.6. *Simulating particle velocity data from strain (rate) data – the spatial integration method*

Olofsson and Martinez (2017) propose a different approach for converting from strain rate to velocity using a spatial integration step. Rearranging equation 3.9, we have

$$\frac{\partial V_z}{\partial z} = \dot{\epsilon}_{ZZ}. \tag{3.18}$$

Next, we perform a spatial integration over z, giving

$$V_z = \int \frac{\partial V_z}{\partial z} dz = \int \dot{\epsilon}_{zz} dz. \tag{3.19}$$

We will call this equation the spatial integration method. The significant improvement of this method over equation 3.14 is the automatic handling of waves of all types without having to separately determine each of their velocities and correct for them individually. Thus, the strain rate data is numerically integrated over depth.

However, as before, this method does nothing to remove the sinc function and still is lacking the geophone response function (in equation 3.12).

Let's look at how this method performs on a synthetic example. Figure 3.4a shows a 400 m depth range of V_z data where the model uses a 70 Hz wavelet propagating at 2500 m/s. Figure 3.4b shows the simulated DAS strain rate data for a 5 m gauge. Figure 3.4c shows the V_z estimated data by performing a depth integration of the strain rate data. Overall, the down-going event seems to have been reconstructed with the exception of a horizontal artifact that starts at the shallow depth and extends across the record. This artifact is due to the limited integration depth range. In the integration process, there is nothing to "cancel out" the shallowest trace, so it persists in the estimate of the V_z data. It is straightforward on synthetic data to remove this artifact using a median filter for each time sample that spans across all traces in depth. The median filtered result is shown in Figure 3.4d. As a practical matter, on field data it is not as easy to remove these horizontal striping events.

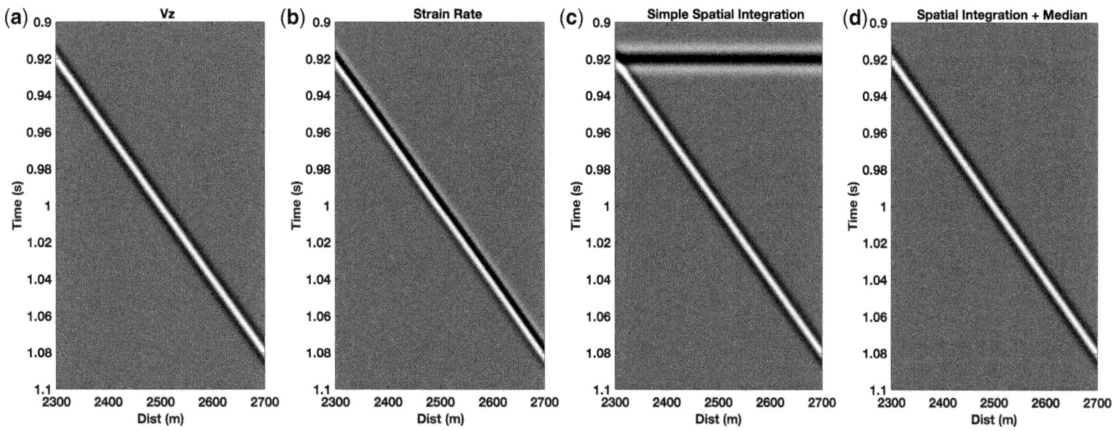

Figure 3.4. (a) The V_z data, (b) DAS strain rate data with a 5 m gauge, (c) strain rate data converted to V_z using simple spatial integration, and (d) strain rate data converted to V_z plus median filter.

Let's look at single traces from the center of this depth zone, at a depth of 2500 m. Similar to the temporal integration method from the previous section, Figure 3.5 shows the results of converting DAS strain rate data to V_z data using the spatial integration method. After median filtering of the converted traces from DAS strain rate to V_z, the converted traces are virtually identical to the results of the temporal integration method.

Without performing median filtering, any event present at the edges of the integration depth window will persist across the resulting converted data. The temporal integration method also performs an integration step, but it is over time. The advantage of integration over time is that we expect there to be no signal at time zero, and we also expect that there will be a "dying out" of the signal at late times. Even if there are signals in each trace in those time windows, we can mute or taper them to reduce these artifacts. However, we usually are averse to muting or tapering the edges of the DAS record in depth. There may be important events in those parts of the record.

Both conversion methods require special handling of the data. For the temporal integration method, the apparent velocity of each event must be determined and its amplitude adjusted after the integration step. In contrast, the spatial integration method requires that the shallow edge of the seismic record be tapered before the integration. Alternatively, a median filter needs to be applied after conversion.

3.7. Simulating particle velocity data from strain (rate) data – comparing methods

In theory, both conversion methods should produce the same estimate of V_z. However, in practice, they will respond to noise differently and potentially will have different numerical limitations. To address this, let's next compare the results of the two conversion methods using slightly more complicated synthetic data. Because the spatial integration method theoretically requires only two DAS traces, but in practice needs an entire depth range of traces, I'll show the result for the spatial integration method first, and then show a single trace comparison of the two methods.

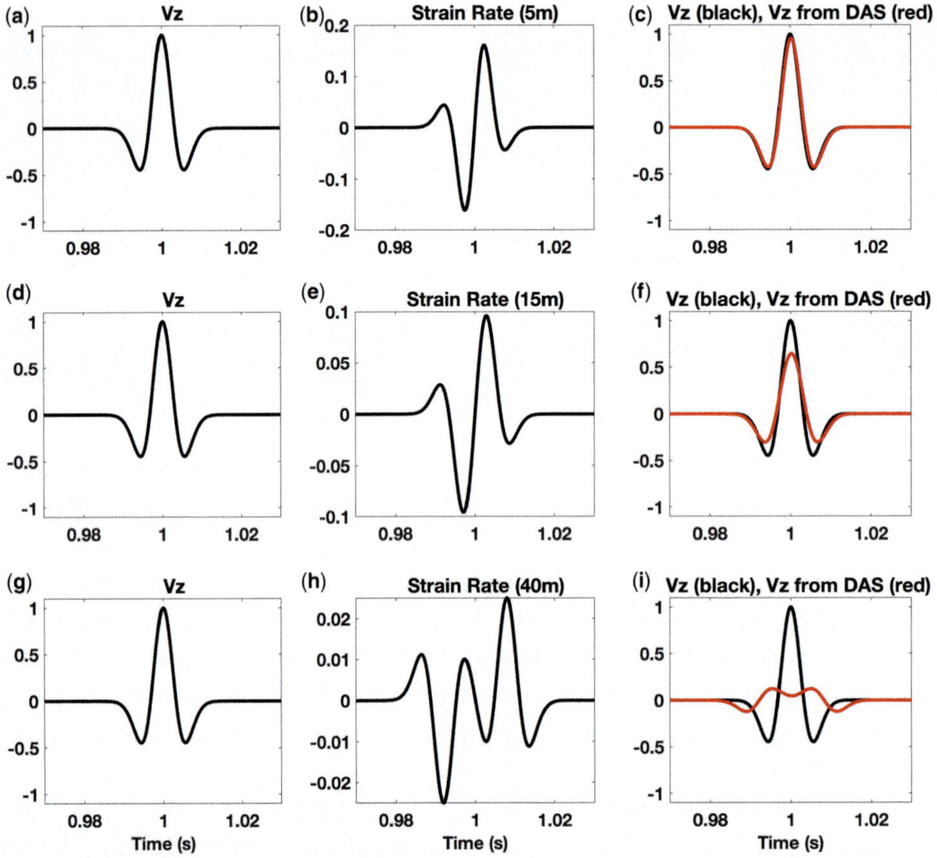

Figure 3.5. Conversion from strain rate to V_z using the spatial integration plus median filter method. Top row: gauge length 5 m. Middle row: gauge length 15 m. Bottom row: gauge length 40 m.

Figure 3.6a shows a V_z record for a model with an interface at a depth of 2500 m. The top layer has a velocity of 2500 m/s, while the lower layer has a velocity of 3500 m/s. An up-going reflection is evident, which starts at $z = 2500$ m and $t = 1$ s. Figure 3.6b shows the DAS strain rate data for the same model. Figure 3.6c shows the result of converting the DAS strain rate data (in Figure 3.6b) to particle velocity, V_z, using the spatial integration method. Next, let's compare how the two methods handle the conversion from strain rate to particle velocity, V_z, at two distinct depths.

The top row of Figure 3.7 shows traces from a depth of 2400 m – this depth is highlighted by the magenta line in Figure 3.6a. The bottom row shows traces from a depth of 2600 m in the model – this is highlighted by the green line in Figure 3.6a. The first column in Figure 3.7 shows magnified areas around the events at depths of 2400 m (Figure 3.7a) and 2600 m (Figure 3.7e). The corresponding DAS strain rate data are shown in the second column for depths of 2400 m (Figure 3.7b) and 2600 m (Figure 3.7f). The corresponding DAS strain rate data converted using the temporal integration and spatial integration plus median filtering are shown in the third and fourth columns, respectively.

Let's start by examining the results for temporal integration in the third column. For this method, we must select the sign of the conversion and the apparent velocity of

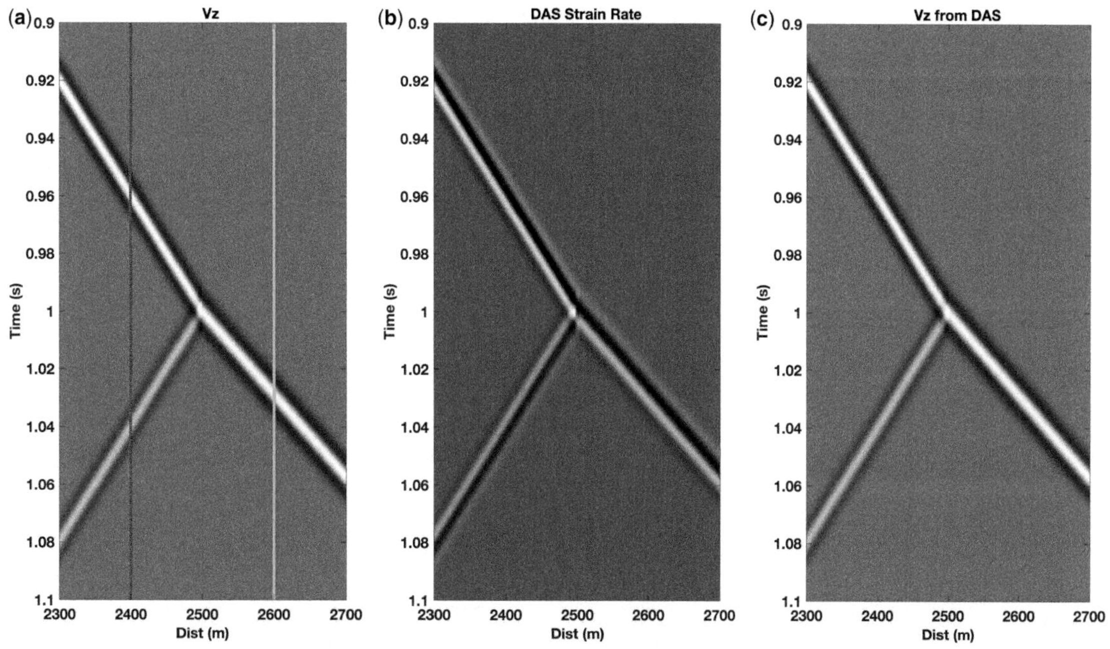

Figure 3.6. (a) Synthetic V_z data for a two-layer model. (b) The DAS strain rate data for the two-layer model. (c) The V_z data converted from DAS strain rate data using the spatial integration plus median filter method.

the single event we wish to convert. We choose to convert the down-going first event on the trace at depth 2400 m. Its apparent velocity is 2500 m/s and its polarity is positive. Figure 3.7c shows that the first event has been properly converted (red) both in sign, shape, and amplitude when compared to the V_z data (black). I have not tried to isolate the second up-going event and convert it separately in the process. We see that using the conversion parameters for the down-going event upon the up-going event has merely flipped its polarity. But because the velocities of the up- and down-going events are the same (ignoring the sign), the wave shape and amplitude of the up-going event is valid.

Figure 3.7g shows the result of using the velocity of the upper layer on the event propagating in the lower layer at a depth of 2600 m. Because this event is also a down-going event, the sign of the conversion is correct. However, the lower layer has a velocity of 3500 m/s, and for the conversion I have used the upper layer velocity of 2500 m/s. We clearly see that the converted data (red) is now too small compared to the actual V_z trace. In fact, it is about 29% ($\approx$ 2500/3500) lower in amplitude than what it should be. Thus, to use this approach, an accurate apparent velocity for each time sample must be computed and applied appropriately or large amplitude errors will be introduced. For complex wavefields, this can be a challenging endeavor.

Finally, let's look at the spatial integration method. There are no parameters to choose for this method, so its application is easy. We see in Figure 3.7d and 3.7h that the converted data (red) match the V_z data (black) nearly perfectly in shape, polarity, and amplitude. It might be noted that there are small artifacts at t = 1.0 s and t = 1.06 s on both spatial integration records (red lines in Figure 3.7d and 3.7h) which correspond to side effects of the median filtering process.

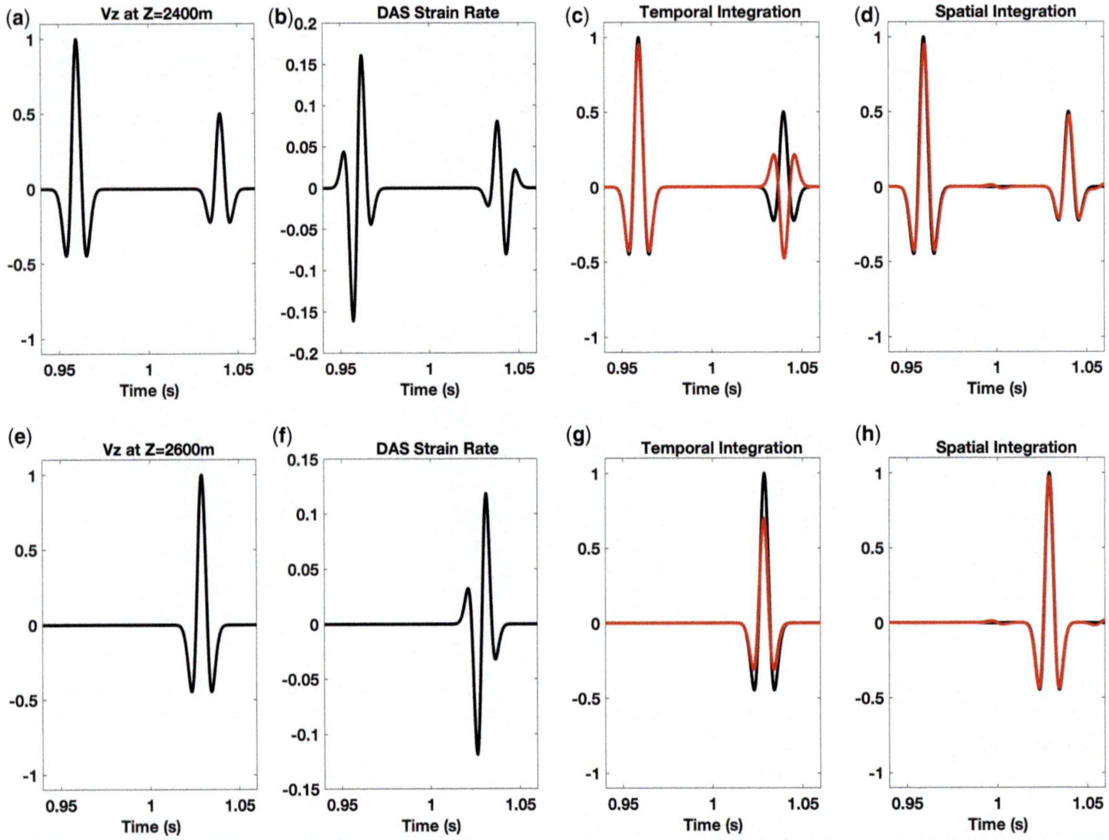

Figure 3.7. Comparison of two conversion methods. The top and bottom rows show traces from depths of 2400 and 2600 m, respectively. The first column shows V_z data, the second column shows DAS strain rate data, and the third and fourth columns show DAS strain rate data converted to V_z using two different methods.

Some of the practical aspects with using the temporal integration will be:

1. The requirement to isolate and convert each event separately with a polarity (sign) and velocity term.
2. The difficulty of handling different events that arrive at the same time and depth. For VSP, examples of this issue will be: tube waves and reflected events arriving at the same time, down-going P-wave multiples, and reflected events.
3. When starting with strain rate data, the temporal integration step will spread out noise spikes (e.g., optical fading covered in Chapter 8) and can cause non-physical DC shifts in the traces.

Some of the practical aspects with using the spatial integration method will be:

1. The simplicity of the method due to the lack of parameters that need to be selected.
2. The need for traces that are closely spaced in depth to perform the integration. To save storage space in some acquisition programs, the DAS data is spatially (trace) decimated from the conventional native 1 m sampling to perhaps 8 m or larger depending on the purpose of the survey.

3. Noise on the traces (from sources, e.g., fading covered in Chapter 8) and inflection points in the arrival times of the events will cause noise "streaks" along the depth direction in the converted data that must be mitigated by a process such as median filtering.

3.8. Simulating geophone data from strain (rate) data

To approximate the geophone function on the DAS strain data converted to V_z, all that is left to do is to apply the geophone response, $R(f)$, as defined in equation 3.12, to the conversion factors in equations 3.16 and 3.17 to yield

$$G_z = \pm c\,\mathcal{F}^{-1}\{\mathcal{F}(\epsilon)R(f)\}, \tag{3.20}$$

$$G_z = \pm c\,\mathcal{F}^{-1}\left\{\mathcal{F}\left(\int \dot{\epsilon}dt\right)R(f)\right\}. \tag{3.21}$$

The parameterization of the geophone function, $R(f)$, in the frequency domain has forced these two equations to implement the operation of the strain term convolved with the geophone response, which occurs in the frequency domain, thus $\mathcal{F}$ and $\mathcal{F}^{-1}$ denote the forward and inverse Fourier transforms in a manner similar to equations 3.13 and 3.15. This makes it seem more complicated than it actually is.

Let's create an example of this process. To minimize the effect of the gauge itself, I'll use a relatively small gauge length of 5 m. The apparent velocity is 2500 m/s and the distance to the DAS channel is 2500 m. To emphasize the effect of the geophone response, $R(f)$, on the signals, I'll use a source center frequency of 20 Hz so that the

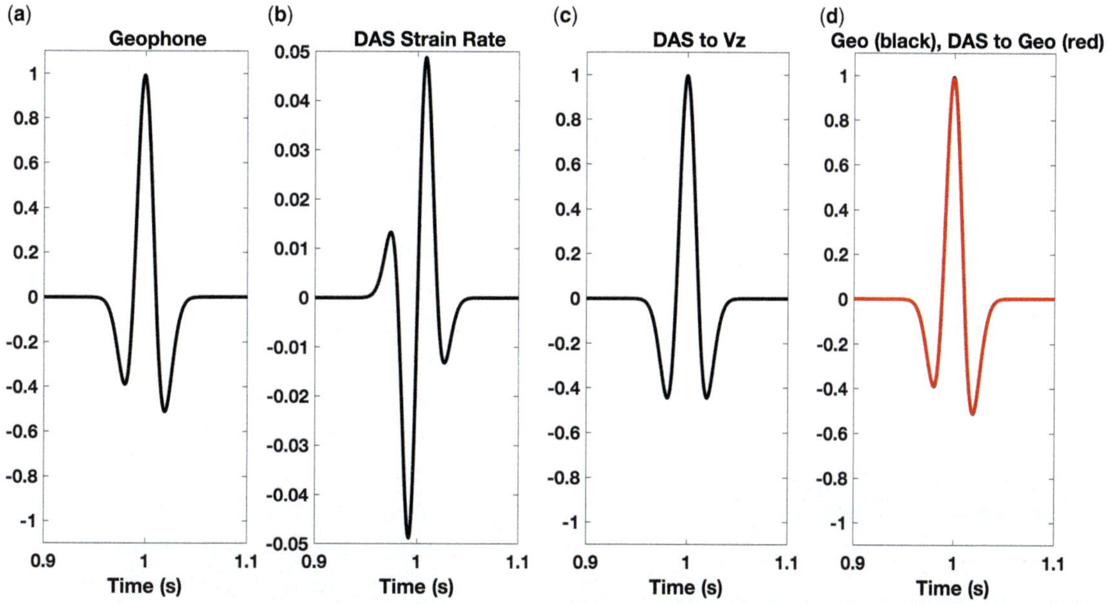

Figure 3.8. Effect of the geophone response. (a) A geophone velocity trace, (b) corresponding DAS strain rate trace, (c) DAS trace converted to V_z, and (d) DAS trace converted to mimic the geophone response (red) plotted over the geophone trace (black).

wavelet is closer to the roll-over point at about 3 Hz in the geophone response curve in Figure 3.2. The geophone trace (after applying the geophone response, $R(f)$, to the V_z trace) is shown in Figure 3.8a, the DAS strain rate trace in Figure 3.8b, the DAS strain rate trace converted to particle velocity in Figure 3.8c, and the geophone trace (black) and DAS strain rate trace converted to geophone response (red) in Figure 3.8d.

Examining the geophone trace in Figure 3.8a, it is clear that the effect of the geophone has been to distort the Ricker wavelet so that the side lobes around the main wavelet peak are no longer symmetric. The side lobe to the left of (or earlier than) the peak is smaller, and the side lobe to the right of (or later than) the peak is larger. This asymmetry is caused by the phase response of the geophone response function in equation 3.12. Because the DAS measurement does not have the geophone instrument response, the extracted DAS V_z trace in Figure 3.8c is a perfectly symmetrical Ricker wavelet. When we next convolve the DAS V_z trace with the geophone instrument response, the wavelet now matches the geophone trace.

3.9. Conversion of phase to a seismic measurement

As I mentioned in Chapter 2, the native measurement that the IU makes is the phase angle between the backscattered light coming from two points on the fiber separated by the gauge length. For clarity, let's change $d\phi$ to θ in equation 2.2 here, so that strain is converted from the phase angle, θ, using

$$\epsilon = \frac{\lambda\theta}{4\pi n GL \xi}, \tag{3.22}$$

where ϵ is strain, λ is operational optical (vacuum) wavelength of the DAS system, n is the refractive index of the sensing fiber, GL is the gauge length, ξ is the photo-elastic scaling factor for longitudinal strain in isotropic material (≈ 0.78), and θ is the phase angle difference of the light measured in radians.

In order to obtain the phase difference measurement, the direct path of the backscattered light is combined with the delayed backscattered light in the interferometer, as shown by the red oval in Figure 3.9. The outputs of the interferometer are two optical signals, I and Q. The I signal is simply the light that comes by directly adding together the light streams. The Q signal is optically modified to be the equivalent of the Hilbert transform of the I signal.

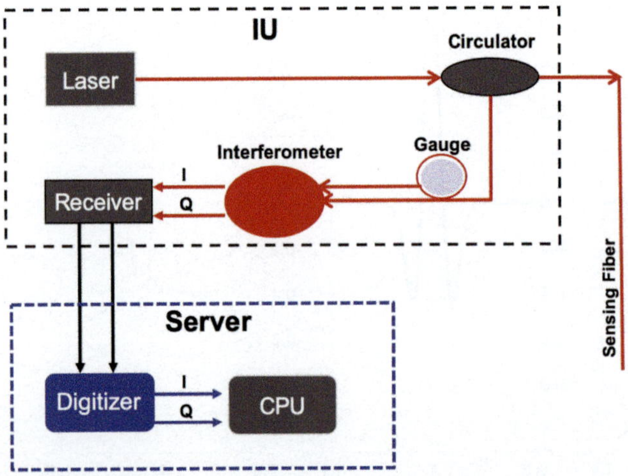

Figure 3.9. Schematic of IU showing details of the optical interferometry step inside the IU (dashed black box) and being digitized in the compute server (dashed blue box).

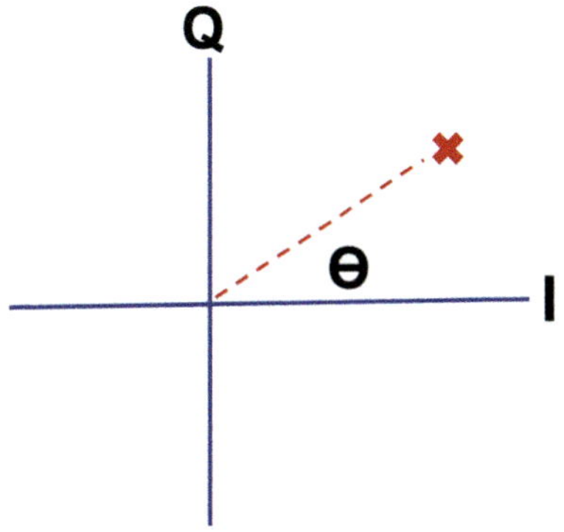

Figure 3.10. Diagram showing how the phase angle, θ, is related to the I and Q values.

Then, these two signals are fed into a photodiode which converts the optical signals to electrical signals. These electrical signals are fed out of the IU and into a two-channel digitizer inside the compute server (shown as the blue dashed box) for the IU system (shown as the black dashed box).

The IU converts the backscattered light to an electrical signal, which then is digitized inside the compute server. This still doesn't seem to give us the phase difference. So, the next step is to form the arctangent of the I and Q signals. For geophysicists, this will remind us that we compute the instantaneous phase of a signal by taking the Hilbert transform and then take the arctangent of the real and imaginary parts. For the I and Q signals, we find the phase angle, shown in Figure 3.10, from

$$\theta = \text{atan}\left(\frac{Q}{I}\right). \tag{3.23}$$

In theory, this is perfect. In practice, we recall that the arctangent returns the angle between $-\pi/2$ and $\pi/2$. That works as long as the phase angle is always between those endpoints. However, we know that, for seismic signals, the phase angle will change by much more than that. Obviously, the computed phase angle from the arctangent must be unwrapped to be continuous. Provided that we have sampled the I and Q signals finely enough in time, algorithms such as that given by Tribolet (1977) will be effective at constructing a continuous phase angle from equation 3.23.

Now that we have the unwrapped phase, we use equation 3.22 to convert the phase to relative strain. By applying the formulas I presented earlier in this chapter, we can go back and forth between relative strain, strain rate, velocity, and geophone response. This is illustrated in Figure 3.11.

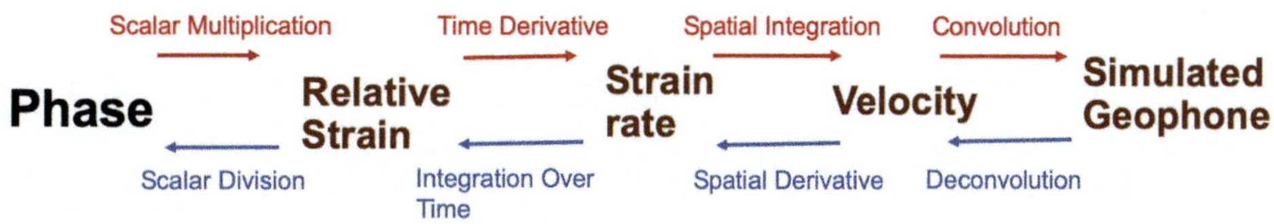

Figure 3.11. Diagram showing the steps to go from phase to simulated geophone response and back. (From Willis et al., 2022.)

You can practice converting phase data to strain and strain rate, and onward to particle velocity and the geophone response using a MATLAB gui (Willis, 2022d).

3.10. Summary

In this chapter, I've presented equations which allow the transformation of DAS strain and strain rate data to the equivalent particle velocity (V_z) and geophone response. I've also presented the inverse allowing the conversion of geophone and particle velocity data to DAS strain and strain rate data. It should be noted that these are idealized equations which assume that we have point-localized measurements of strain. This is definitely not the case for DAS data. The DAS strain and strain rate measurements are made over a finite distance – over the gauge length. Thus, DAS makes a good approximation of the strain and strain rate over a given frequency bandwidth controlled by the gauge length and the apparent velocity of events being recorded. Finally, I have presented how the phase measurement is converted into the strain (rate) values and graphically have shown the relationships between all of these measurements.

Chapter 4 Fiber optic cables

If I were not a physicist, I would probably be a musician. I often think in music. I live my daydreams in music. I see my life in terms of music.

– Albert Einstein

Comme c'est curieux! Comme c'est bizarre! Et quelle coincidence!

– Ionesco's La contratrice chauve

4. Types of optical fibers

Fiber optic cables are used in many data transfer applications such as telecommunications as well as sensing applications including DAS, distributed temperature sensing (DTS), and distributed temperature and strain sensing (DTSS). In this chapter, I will review some of the important aspects of fiber optic cables with respect to these applications.

4.1. Scattering mechanisms of laser light in optical fibers

When laser light is introduced into one end of an optical fiber, the overwhelming portion of the light propagates away from the laser and toward the distant end of the fiber. However, a small portion of the light is scattered by minute defects, material property changes, and/or reflective features distributed throughout the length of the fiber. This scattering occurs in all directions by each defect. However, because the fiber is constructed to only propagate light along its longitudinal direction, only forward scattering can propagate in the direction away from the laser, and backscattering can propagate in the direction back toward the laser. All of the other directions are attenuated or radiated away by the construction geometry of the fiber and the destructive interference of the scattered energy.

Characteristics of the scattered light can be put into the three primary categories, shown by the illustration in Figure 4.1, as Rayleigh, Brillouin, and Raman. Some sources show illustrations of this spectrum as a function of wavelengths like I've presented in Figures 4.1 and 4.2 (Baihly et al., 2015, Figure 3; Amira et al., 2015, Figure 2), while others show it as a function of frequency (Thévenaz, 2011, Figure 9.2). So you need to be careful when viewing these illustrations because, as will become clear, the Stokes and anti-Stokes side of the illuminating laser light will flip between the two different representations.

For many applications, the vacuum or free-space wavelength of the incident laser light is about 1550 nm, so I've used that wavelength to construct this example.

Note that optical wavelength is compressed or reduced when propagating inside the optical fiber due to its reduced velocity compared with free-space velocity of approximately 3×10^8 m/s. The velocity of light in the fiber is reduced due to the increased density of the optical media compared with vacuum or air. Inside the fiber, light travels at closer to 2×10^8 m/s (about 1/3 slower). The index of refraction (IOR) or refractive index is the unitless ratio of the velocity of light in a vacuum to its velocity in the optical medium (i.e., glass). As such, if scattering is an elastic process, the electromagnetic frequency of oscillation does not change; however, in all cases, the wavelength of light decreases inversely proportional to the index of refraction.

Rayleigh scattering (red lines in Figures 4.1 and 4.2) occurs at the same wavelength (or frequency) as the incident light and, thus, is an elastic process. Both Brillouin and Raman scattering processes are inelastic by nature, meaning that new frequencies of light are generated via nonlinear light-matter interaction.

Using some typical average material phonon values at room temperature for silica optical fiber, the shift in optical wavelength of Brillouin (blue in Figures 4.1 and 4.2) and Raman (magenta in Figure 4.1) scattered waves is 0.09 nm (about 11 GHz) and 96 nm (about 13THz), respectively, at an interrogation wavelength of approximately 1550 nm (Thévenaz, 2011, Chapter 9). Thus, the change of wavelength for Brillouin

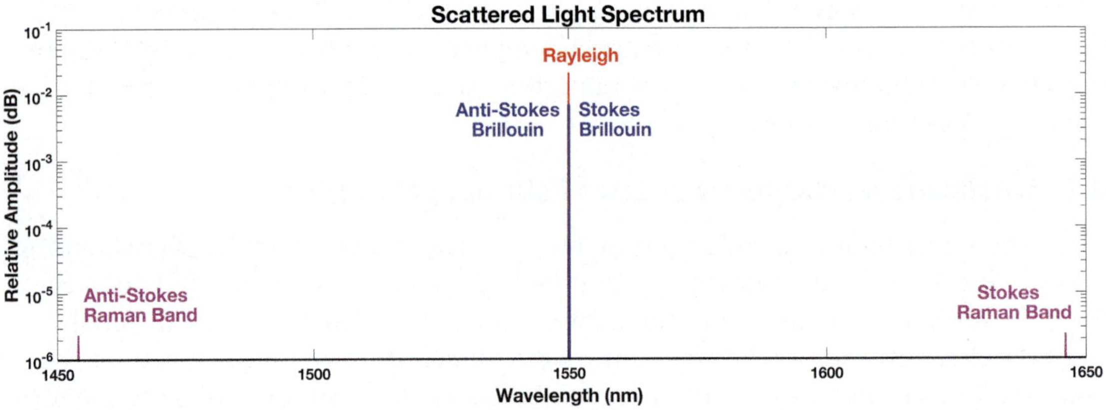

Figure 4.1. Fiber optic scattering mechanisms.

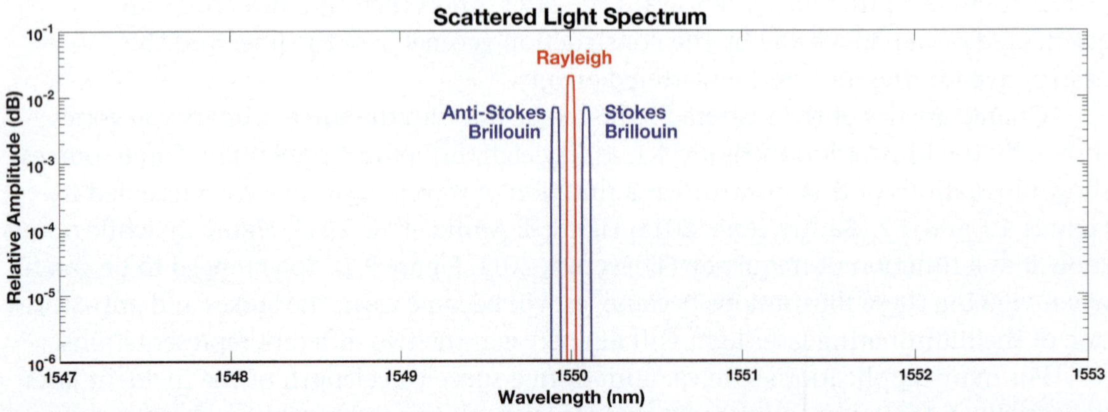

Figure 4.2. Magnification of spectrum shown in Figure 4.1.

scattered waves is proportionally small, compared to the much larger change for Raman scattered waves. Figure 4.2 is a magnification of Figure 4.1 so that the wavelength shift for the Brillouin waves can be seen. The notional amplitude differences shown between the three scattered wave types indicate that, by far, Rayleigh scattering is the strongest, while Raman scattering is the weakest.

4.1.1. Rayleigh scattering (elastic scattering)

Distributed acoustic sensing utilizes Rayleigh backscattered light (or more specifically coherent Rayleigh backscatter) to detect and measure acoustic signals incident on the fiber. An overview of this process is given in Section 2.2 and will be covered in more detail in Chapters 5, 6, and 7. Of the three scattering mechanisms, Rayleigh scattered light is probably the easiest to understand. When light interacts with particles in the fiber that are less than about 1/15th of its reduced wavelength in the fiber, light will be scattered by what is called Rayleigh scattering, named after Lord Rayleigh (born John William Strutt) who discovered it (Hecht, 1998). This scattered light will have the same wavelength (or frequency) as the incident light. Forward Rayleigh scattered waves always add together in phase but have a phase lag from the propagating incident light (Thévenaz, 2011, section 8.1.1). This means that the light reaching the far end of the fiber will be modified in amplitude and phase by the interaction of the forward scattering waves.

Rayleigh scattering gives rise to wavelength-dependent attenuation in ordinary optical fibers, where attenuation is inversely proportional to the 4th power of the wavelength ($\sim 1/\lambda^4$). For wavelengths less than approximately 1600 nm, Rayleigh scattering is the major cause of intrinsic optical attenuation in optical fibers. Typical DAS operational wavelengths are near 1550 nm. The total optical losses for a wavelength of 1550 nm in a typical optical fiber is about 0.2 dB/km (about 5%/km). For a cable of 15 km in length, this translates to losses due to optical attenuation of about 3 dB. Rayleigh scattering contributes most of these losses, in fact about 0.15 dB/km (Guenot, 2003) for a total of 2.25 dB in 15 km of fiber. The actual power of the Rayleigh backscattered energy is also dependent upon the construction of the fiber (see, for example, Thévenaz, 2011, equation 8.25).

4.1.2. Brillouin and Raman scattering (inelastic scattering)

We have seen that, in Rayleigh scattering, the scattered waves have the same frequency or wavelength as the incident light. However, in both Brillouin and Raman scattering, the wavelength is changed by the nonlinear interaction of the light and silica material of the fiber. For both of the mechanisms, two scattered waves are created with wavelengths shifted by a symmetric amount about the wavelength of the incident light. (Actually, it is symmetric on the frequency axis, not the wavelength axis.) The scattered light with smaller (shorter) and higher (longer) wavelengths is called anti-Stokes and Stokes scattered waves, respectively.

The incident light in both of these mechanisms interacts with the molecular glass matrix to generate optical phonons (Raman) and acoustic phonons (Brillouin) in the fiber, where a phonon is simply an arrangement of atoms or molecules that are collectively vibrating together. The background thermal activity in the fiber provides phonons in each possible vibration state given by the Bose-Einstein statistical distribution (Thévenaz, 2011, Section 9.2).

In anti-Stokes scattering (Figure 4.3a), the incident light photon interacts with the scattering site phonon vibration and is annihilated. For conservation of energy, a new scattered photon of light is created at a *shorter* wavelength (or higher frequency, and higher photon energy). At the same instant, the phonon in the fiber that is *annihilated* has the corresponding resonant frequency of the difference between the incident and inelastic scattered waves.

In Stokes scattering (Figure 4.3b), the incident light photon interacts again with the scattering site phonon vibration and is annihilated, but in this second scenario the scattered photon of light created has a *longer* wavelength (or lower frequency, and lower photon energy). Also, for conservation of energy, in this second scenario, a new phonon is *created* which resonates at the frequency difference between the incident and scattered waves.

4.1.3. Brillouin scattering

As mentioned, Brillouin backscattered light is typically lower in amplitude than Rayleigh backscattered light and relatively close to the wavelength of the incident light within about 11 GHz at 193.1 THz optical frequency, assuming a vacuum wavelength of about 1550 nm, as can be seen in Figure 4.2. Brillouin scattering will produce anti-Stokes and Stokes contributions, in about equal amplitudes (Thévenaz, 2011, Section 9.2). Brillouin scattering is dependent upon both the temperature and strain of the fiber. If the temperature is known at all points along the fiber, then the strain can be extracted accurately. This process is known as distributed temperature and strain sensing (DTSS). The attraction of this method is that Brillouin scattering is much higher in amplitude than Raman scattering, and thus easier to measure.

Raman-based signal processing typically involves using a pair of optical filters and photodetectors to separately measure the ratio of anti-Stokes to Stokes light intensities. A disadvantage is that the intensity levels of both signals can vary due to differential attenuation in the fiber at each wavelength. Raman systems also need to be regularly recalibrated against temperature measurement drift.

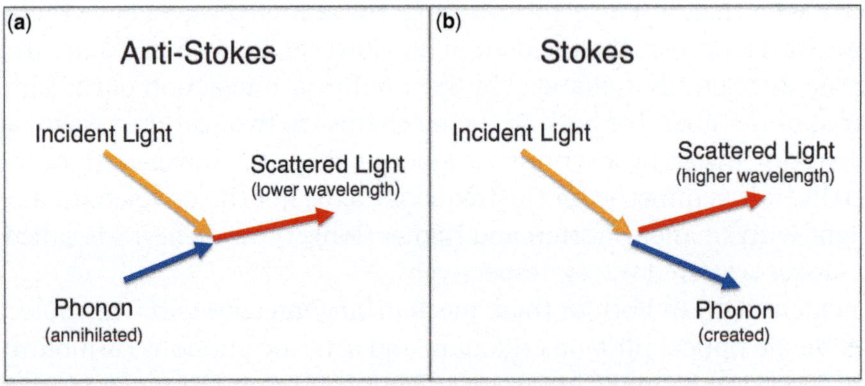

Figure 4.3. Schematic showing (a) the anti-Stokes scattering and (b) the Stokes scattering.

As for Brillouin-based signal processing, the disadvantage is that, because the signals are close in wavelength to the original interrogation wavelength, Rayleigh scattering can be difficult to separate from anti-Stokes and/or Stokes light using optical filters. As such, special coherent detection methods that use radar-style frequency demodulation techniques are employed to extract temperature- and/or strain-induced Brillouin acoustic phonon frequency shift, which represents the change in temperature and/or strain along the fiber. Note that both single-mode and multi-mode fibers can be interrogated with Brillouin-based instruments, but single-mode fibers will typically always give better optical signal-to-noise ratios resulting in the highest possible temperature-strain measurement accuracy and spatial resolution (typically 1–3 m at tens of km distance).

4.1.4. Raman scattering

The wavelengths of the backscattered anti-Stokes and Stokes Raman contributions are much farther away from the wavelength of the incident light than Brillouin scattering. The Raman optical backscatter cross section is smaller in amplitude than the other two backscatter mechanisms. The anti-Stokes Raman scattering will be approximately 8 times smaller in amplitude than the Stokes Raman contribution (Thévenaz, 2011, Section 9.2). The anti-Stokes Raman scattered waves also will be strongly dependent upon temperature, while the Stokes Raman scattered waves will be much less sensitive to temperature. The ratio of the light intensities of the anti-Stokes Raman to Stokes Raman scattered light can be used to extract temperature all along the fiber (Bolognini et al., 2007). This process is called distributed temperature sensing (DTS).

4.2. Single-mode and multi-mode fiber

A typical optical fiber is composed of three essential parts – the silica glass core, the silica glass cladding, and the protective buffer and jacket materials. The core is the glass in the center of the fiber that is used to send light, and thus information, through the fiber. The purpose of the cladding is to create a waveguide or light pipe, trapping the light inside the core glass via total internal reflection. The refractive index of the cladding must be lower than the refractive index of the core glass for total internal reflection to occur. In Section 2.2.1, we cover the criteria for total internal reflection and critical angle of refraction (equation 2.1), which allows trapping and low loss longitudinal propagation of light inside the core.

There are two major types of fiber, as shown in Figure 4.4, used for distributed sensing applications: single-mode fiber (SMF) and multi-mode fiber (MMF). In single-mode fiber, the core glass is typically about 9 μm in diameter, which ensures that essentially only one "straight" path or "mode" of light can propagate through the fiber.

[Interesting to note is the fact that single-mode fibers actually propagate two orthogonally degenerate polarization state modes where total power is shared between the two polarization modes. Due to the fact that ordinary single-mode fiber is birefringent, the two polarization states can change their relative phase relationships

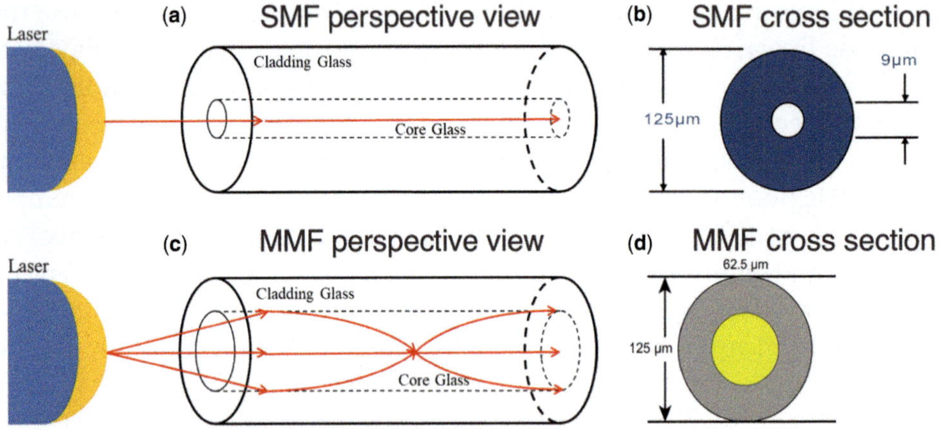

Figure 4.4. (a) Single-mode fiber light path, (b) single-mode fiber cross section, (c) multi-mode fiber light paths, and (d) multi-mode fiber cross section. (From Willis et al., 2018.)

depending on mechanical stress and temperature variation along the fiber. This is especially important in interferometric sensing applications such as DAS.]

In contrast to single-mode fibers, multi-mode telecommunications fibers have a much larger core, typically 50 μm or 62.5 μm. This larger core allows for multipathing of light, or multiple modes of light, to propagate in the fiber. Depending on the optical source, up to several hundred modes may be excited with each carrying a fraction of the total energy or power propagating.

Because the core is larger in a multi-mode fiber, much more light is able to be pumped into it. The disadvantage of multi-mode fiber is that the multiple modes of light will eventually disperse out and eventually travel at different velocities. Each light pulse will spread out in time (via intermodal distortion or intermodal delay). This limits the effective maximum length (or bandwidth-distance product) of multi-mode fiber. In contrast, while single-mode fiber does not allow as much light to be pumped into it, because the light travels in a single simple path, it can be used over much longer distances. This is because the pulse dispersion or bandwidth-distance product is limited mostly by chromatic dispersion, due to nonzero source spectral linewidth. Laser sources are relatively very narrow, spectrally, and are the dominant optical sources employed in long reach fiber optic sensing systems, due to their modulation speed, optical power coupling efficiency, and low chromatic dispersion characteristics.

Graded-index (GI) multi-mode fiber has been used for many years for distributed temperature sensing applications. In these multi-mode fibers, the index of refraction of the core is parabolically graded to force lower order modes to propagate more slowly compared with higher-order modes, thus minimizing temporal pulse spreading over distance. As such, many older wells will have permanent graded-index multi-mode fiber installed in them. Single-mode fiber is the preferred medium for use in DAS applications. However, it is possible to have a single-mode fiber DAS IU with multi-mode sensing fiber if some additional hardware modifications are made, but the signal-to-noise (S/N) level of the resulting strain rate data will be lower than when using single-mode sensing fiber. Although the signal levels are comparable, Willis et al. (2018) report an S/N reduction of approximately 4 dB from the increased noise with the multi-mode fiber when compared to the single-mode fiber.

4.3. Engineered fiber

There are many different design modifications that can be made to the fiber when it is manufactured.

One change is to reduce the amount of impurities, or scattering sites, in the fiber. This is called ultra-low loss (ULL) fiber. This type of fiber is desirable in the non-sensing portions of the optical path from the IU to the well with sensing fiber (Ellmauthaler et al., 2020b) or to the cable on the ground with sensing fiber in surface applications. This will reduce the optical attenuation in the transmission portion of the fiber and increase the distance the light can propagate with less loss prior to reaching the sensing fiber.

An engineered fiber uses the diametrically opposite approach, which is to increase the amount of backscattering by modifying the fiber using one of several methods. Westbrook et al. (2017, 2020) and Wu et al. (2020) modify a standard single-mode fiber by introducing distributed ultraweak reflective fiber Braggs gratings (FBG) into the fiber. This is done using a continuous ultraviolet laser or femtosecond laser inscription beam and optical phase masks. (It also can be done using a two-beam interference-based photolithographic method to generate the fiber Braggs gratings or reflective features along the fiber.) The result is more than 10 dB of increased backscattered light and less than 0.5 dB/km attenuation. This is a tremendous increase in the backscatter return signal level with only about an extra 0.3 dB/km excess transmission loss, at 1550 nm wavelength, compared with normal optical attenuation as the light propagates.

In choosing the type of fiber to be installed for your application, consider the following points:

1. If you have a fiber optic cable in excess of 5 km, and if there are long portions of that cable which is simply connecting the IU to the desired sensing fiber, and you are not interested in obtaining measurements in that portion, then consider replacing that portion of the cable with ultra-low loss fiber.
2. If you expect low amplitude dynamic strain rate signals, which you need to monitor through long transmission and interconnection lengths and at high sampling rates for acoustic bandwidth requirements, consider using engineered fiber with increased backscattering to improve optical signal-to-noise ratios.

4.4. Summary

Impurities in the optical fiber cause scattering of incident light within the fiber. While there are three types of scattering, Rayleigh scattering produces the largest amount of scattered light and is the dominant mechanism used for DAS measurements today. Rayleigh scattering is an elastic process with scattered light having the same wavelength (and frequency) as the incident light.

The diameter of the glass core determines whether the fiber acts as a single-mode or multi-mode waveguide at optical wavelengths greater than the "cut-off" wavelength, which is typically about 1250 nm. (Some ultra-low loss transmission fibers may have cut-off wavelengths closer to 1500 nm.) An appropriately small core will only allow a single mode of light to propagate and allows for a long distance of propagation without appreciable temporal pulse spreading. A larger glass core will allow more light to be coupled and pumped into it and will create multi-pathing of the light within the core. Raman backscatter-based distributed temperature sensing (DTS) systems typically use multi-mode fiber and, therefore, many older wells have it permanently installed. While it is possible to use multi-mode fiber for DAS measurements, it is preferable to use single-mode fiber as it can acquire higher signal-to-noise ratio data over longer distances and optical transmission loss power budgets.

Ultra-low loss optical fibers, with near-pure fused silica (SiO_2) cores, are specifically designed to reduce the amount of backscattering and thus reduce optical attenuation. These can be used for portions of the optical cabling connecting the IU to the sensing fiber where no strain measurements are needed. Optical fibers can be modified or engineered to create more backscattering and thus better signal-to-noise ratio strain data.

4.5. Acknowledgement

Much of the technical detail in this chapter has been contributed by John Maida, who has deep expertise in the physics of fiber optics. I thank him for reviewing this chapter and making many edits and suggestions.

Chapter 5 Architecture of a DAS acquisition system

What is research but a blind date with knowledge.

– Will Harvey

5. Architecture and components of the DAS interrogator system

It is important to understand the way a DAS acquisition system is designed. The different individual components need to be assembled in an optimal architecture to obtain seismic data with the best possible signal-to-noise ratio (S/N). The IU itself can be implemented in many different manners, e.g., as a homodyne or heterodyne optical system (Hartog, 2017), but in this chapter I will use a homodyne system to describe how the IU interacts with the entire DAS acquisition architecture. I will discuss two basic VSP systems as a proxy for many other possible acquisition configurations.

5.1. Layout of the DAS acquisition system

The DAS acquisition systems can be deployed for many different applications and in many environments. Figure 5.1 shows a schematic deployment for land VSP acquisition. The recording truck is located near a takeout connection for the fiber optic cable. Typically, a short (<100 m) fiber optic jumper cable is connected from the IU to the takeout junction connection for the fiber optic cable surface cable leading to the well. To protect the surface cable from damage by vehicles, foot traffic, and animals, typically it is buried in a shallow trench. Then, the surface cable is connected to the sensing fiber in the well by a dry-tree connector. This connector is constructed to allow the fiber to pass through the well head while sealing it from possible high-pressure fluid leaks of the drilling and/or produced fluids.

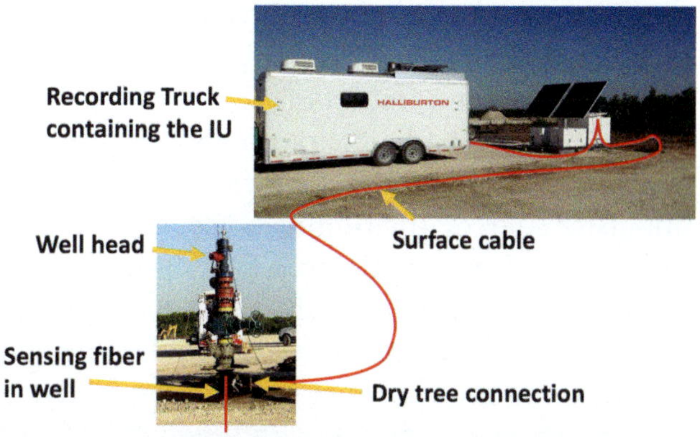

Figure 5.1. Schematic layout of a typical land DAS VSP acquisition system. (Courtesy Halliburton.)

Figure 5.2. Schematic showing an example layout of a subsea DAS VSP acquisition geometry. (Modified from Ellmauthaler et al., 2020b, Figure 1; Wilson et al., 2020b.)

Figure 5.2 shows a much more complicated deep water, subsea deployment for DAS VSP acquisition. The IU is located on a floating production storage and offloading (FPSO) vessel. An umbilical cable connects the IU on the FPSO to the drill centers on the sea floor. The umbilical cable is further connected to the sensing fiber in the well by use of a wet-tree connector. Deploying fiber optic cables in the subsea environment is complicated by the usually very long umbilical cables, and multiple optical connectors are required which create optical losses that must be mitigated.

5.2. Time keeping

An important aspect of acquiring seismic data is knowing what time each recorded trace sample occurred. For earthquake and microseismic recording, precise timing is critical so that the location of events can be determined and matched with the master catalogues of events recorded by other stations.

Timing is also important for active source seismic data sets, such as surface seismic and VSP data. When vibrator sources are used, accurate timing is required so that the correlation (or deconvolution) with the source sweep can be made appropriately. Even for impulsive sources, the timing of the records must be known so that formation properties can be inverted using the true traveltime from source to receiver. For blended surveys, where sources are fired nearly simultaneously with each other, having accurate timing for the records is required in order to deblend the recorded data back into individual shot records using the known global position system (GPS) time of each source (Ikelle, 2010).

In Figure 2.7, I introduce a schematic of a simplified homodyne IU. In Figure 5.3, three fiber stretchers have been added to that original figure. The right side of the figure shows a photo of a fiber stretcher which is a piezoelectric (PZT) spool around which an optical fiber is wound. An electrical signal can energize the PZT, which in turn stretches the optical fiber wound around it and thereby encodes that auxiliary signal into the optical data stream. This example shows that the three fiber stretchers are encoding

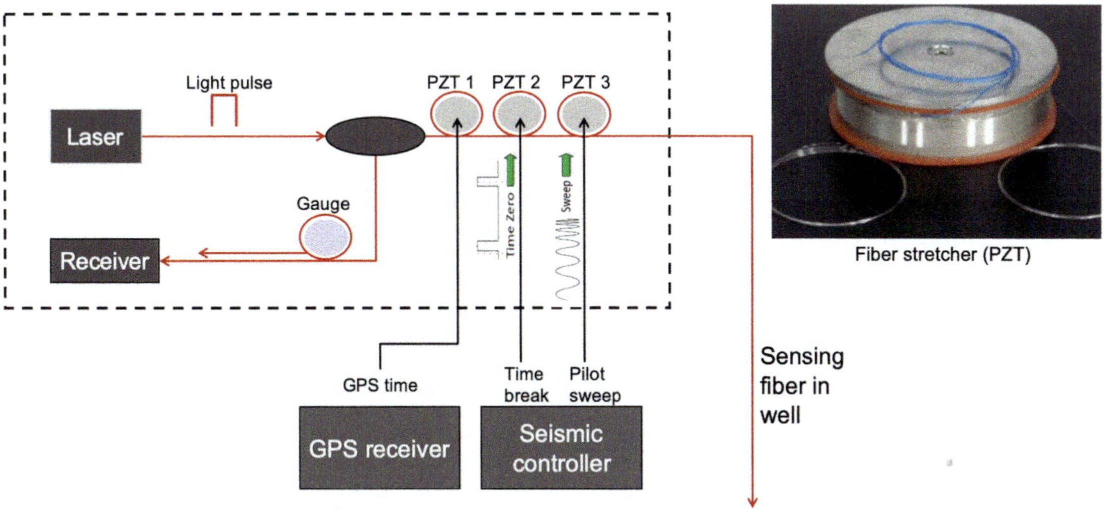

Figure 5.3. Schematic of an IU with three fiber stretchers (left) and a photo of a fiber stretcher (right).

data from the GPS time clock, while the other two fiber stretchers are encoding the time zero pulse and sweep drive signals from the seismic controller.

The IU can be set up to record individual shot records or continuous records. Each option has advantages. To record individual shot records, the start of the recording is triggered from the time zero pulse from the seismic controller. The channel of data coming from the second fiber stretcher, PZT2 in Figure 5.3, is monitored, so that the IU system starts to record the DAS data when a pulse is detected. The value of triggered acquisition is that it is easy to perform quality control (QC) on the data in the field because the individual shot records are exported as SEGY files in real time and then viewed in a seismic processing system.

Alternatively, in the continuous recording mode, the stream of DAS data is recorded to a computer file in uniform blocks of time. The key to deciphering these records of continuous data is to examine the GPS time channel from the fiber stretcher labeled PZT1 in Figure 5.3. For active seismic source surveys, the advantage of this method is that no radio communication between the seismic controller and IU is required, which makes it ideal for large marine surveys and land surveys where radio signals may be blocked by mountains or cultural features. Then, the log of the source firing times recorded elsewhere can be used to separate the continuous DAS data into individual shots records.

5.3. Pulse rate

I've mentioned that pulses of laser light are sent down the fiber and that backscattered light is used in the IU to measure strain in the sensing fiber. From the signal processing sampling theorem (Oppenheim and Schafer, 1975), we know that the finer time sampling we use, the higher frequencies we can measure. Typically, IU systems pulse the laser light into the fiber at between 4 kHz and 20 kHz (Ellmauthaler et al., 2020a). This means that data acquired with a 20 kHz pulse rate will have a spectral content 5 times greater than for acquisition using 4 kHz.

It is not uncommon for the DAS strain signal to be recorded in SEGY records at 1 ms sampling interval. This corresponds to a Nyquist frequency of 500 Hz. So, you must be thinking, why do I want a signal bandwidth of over 500 Hz, when my typical seismic frequencies are usually less than 100 Hz?

The background noise level in fiber strain data is relatively flat at all frequencies. As you recall from basic signal processing, any time we decimate data, we need to perform a low pass filter to avoid aliasing the high-frequency noise back into the lower frequency signal. We record the strain data with a large bandwidth (up to 20 kHz) and then the S/N is increased through field data processing. Next, the data is decimated to 1 ms sampling after applying a low pass filter that removed noise above the Nyquist frequency of 500 Hz. Having the data sampled at up to 20 kHz, the aliasing that occurs when the data is down-sampled is greatly reduced.

Although it is good to pulse the laser light at high rates, there is a limitation on that rate. Each pulse of light needs to travel from the IU, to the end of the sensing fiber, and back to the IU before the next pulse starts. Thus, the timing depends upon the length of the surface and sensing fiber cables and the speed of light in the fiber. Equation 3.4 from Hartog (2017) gives the equation of the time it takes light to travel from the IU to the end of the fiber and back to the IU:

$$dt = {}^{2}\!/_{v_g}\, dz, \tag{5.1}$$

where dt is the two-way traveltime, v_g is the group velocity of light in the fiber, and dz is the length of the fiber. The group velocity of light in the fiber is given by (Alfataierge et al., 2020, equation 1)

$$v_g = C/n, \tag{5.2}$$

where C is the speed of light in a vacuum, and n is the fiber refractive index. Using $C = 299{,}792{,}458$ m/s and a typical value of $n = 1.4679$ (Corning, 2021), the result is $v_g = 196{,}972{,}705$ m/s.

Figure 5.4 shows a graphical representation of equation 5.1, in black, using these constants. This provides an estimate of the maximum pulse rate allowed in the fiber versus the length of the fiber optic cable. For reference, the red dashed line is the sampling frequency (1 kHz) for seismic data acquired at a 1 ms sampling interval. From this graph, you might be tempted to say that we can acquire seismic data all the way out to 100 km. However, the S/N of the data will be reduced because the acquisition sampling should be much higher than 1 kHz in order for the anti-aliasing filters to work properly.

When the DAS strain (rate) data is used for monitoring the injection or production flow noise in pipes, the noise spectrum can easily exceed 4 kHz (Carpenter, 2016; Stajanca et al., 2018). In order to accurately acquire the flow noise data, the sampling interval should exceed twice the upper limit of the frequencies we need to acquire. If the upper desired frequency band is 4 kHz, for example, the sampling interval would need to be 8 kHz or higher, as shown by the blue dotted line in Figure 5.4. This means that fiber optic cables longer than 12 km would not be able to accurately sample the

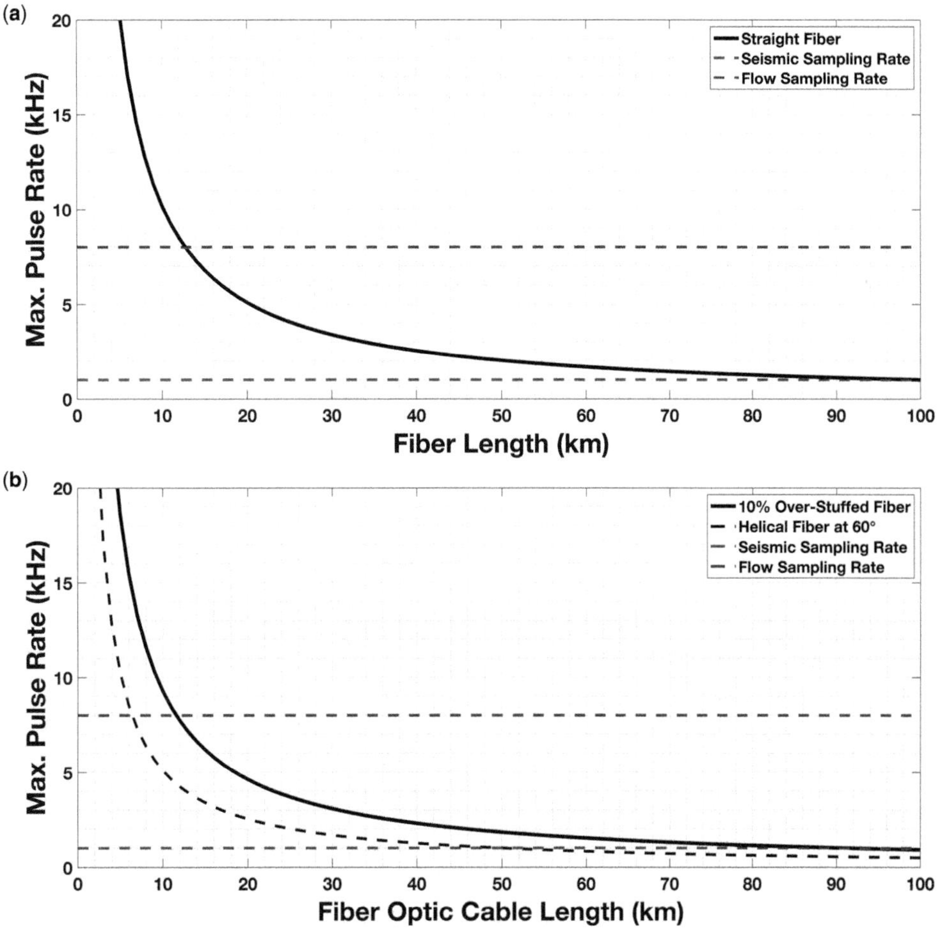

Figure 5.4. (a) Relationship between the maximum pulse rate versus fiber length (black). The red and blue dashed lines show the minimum (but not optimum) sampling frequency for acquiring typical seismic data and flow noise, respectively. (b) Relationship between the maximum pulse rate versus fiber optic cable length for two types of optical fiber geometries.

strain (rate) signal. Optimally, the cable should be less than 10 km because, for this distance range, the anti-alias filter can help attenuate the higher noise frequencies which would otherwise alias back into the signal, which is called wrap-around noise.

The graph in Figure 5.4a assumes that the fiber in the cable is actually the same length as the cable itself. However, when fiber optic cables are constructed, the optical fiber is intentionally "over-stuffed," so that when the entire cable is put under tension, there is no likelihood of the optical fiber breaking. The mechanical strength and stretching properties of the entire cable will be very different from the optical fiber itself (Ander et al., 1990; Coulter and Goodman, 1991; World Oil, 2008; Corning, 2021). For optical fiber that is simply run "straight" through the cable, it will typically be over-stuffed by about 10%.

There are other fiber optic cables constructed so as to wind the optical fiber helically around a central mandrill in the cable to change the sensitivity of the fiber to seismic energy, which will be discussed in Sections 7.3 and 7.4. These configurations will greatly increase the length of fiber inside the cable and thus reduce the maximum pulse rate. Figure 5.4b shows the maximum pulse rate versus the fiber optic cable length with 10% over-stuffed fiber (black) and helically wound fiber at 60°(black dashes). While over-stuffing slightly decreases the maximum pulse rate, helically wound fiber greatly reduces it.

Better S/N is always a goal, but one question is how much is enough? Of course, this depends on the application and the data acquisition geometry and parameters. For a simple zero offset VSP, my model-based studies have shown that, for extracting the velocity profile from the first breaks, an S/N of 26 dB is required. For creating a corridor stack, this drops to 16 dB. These levels can be obtained by acquiring multiple records by repeating the sources, using larger energy level sources, and adding signal processing.

5.4. Extending the range of DAS acquisition

Fiber optic cables for long haul telecommunications have been in use for more than 30 years. For example, the first intercontinental fiber optic cable run on the Atlantic Ocean sea floor dates back to 1988. With the FLAG (Fiberoptic Link Around the Globe) network stretching 28,000 km from the United Kingdom to Japan (Verizon News Archives, 1997), it might be counterintuitive to think we need to be concerned about the length we can deploy a fiber optic cable to obtain DAS data. The difference is that for DAS we need to detect and use the comparatively very small backscattered light signals, not the forward propagating light. Plus, we have to wait for the light to travel to the end of the fiber and back to the IU. Those are our primary challenges. In addition, the topology of the fiber cable path adds many additional losses.

Figure 5.2 shows a topology for a subsea DAS acquisition system. Willis et al. (2021) show the additional expected losses from typical subsea DAS acquisition topologies compared with a simple land case, as shown in Table 5.1. The dry- and wet-tree labels refer to the way the fiber connects at the well head. The subsea topology additional losses are annotated in Figure 5.5.

Several optical connections need to made, connecting the umbilicals to the drill centers, the drill centers to the well head, and the sensing fiber to the well head. There are two basic types of optical connections made – dry-mate and wet-mate. As the name implies, a dry-mate connection is made manually under dry conditions, and because of that care can be taken to ensure a low loss connection. Alternatively, wet-mate connectors are created for splices that need to be made subsea, typically by robotic

Table 5.1. Typical causes for optical losses for land and subsea DAS systems; the S/N loss column assumes 10 km of umbilical fiber and eight wet-mate connectors plus losses at the wet-tree.

Source of optical loss	Dry-tree (land)	Wet-tree (subsea)	S/N loss
Down hole sensing fiber length	≤10 km	≤10 km	–
Umbilical fiber length	–	5 km–100 km	3.4 dB
Pulse repetition rate	10 kHz–50 kHz	1 kHz–8 kHz	3 dB
Optical connections (dry/wet)	1/–	≥2/≥2	15 dB
Total expected optical losses			**21.4 dB**

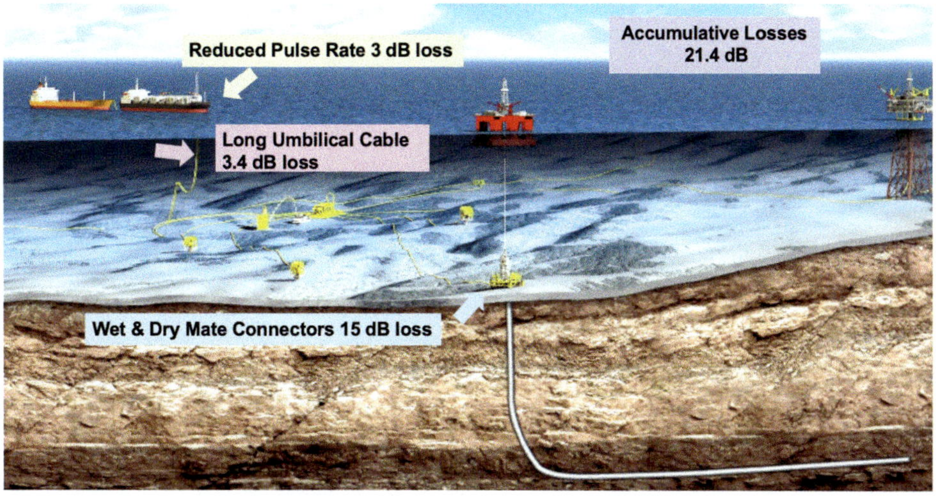

Figure 5.5. Typical optical losses from a subsea DAS acquisition deployment. (From Willis et al., 2021.)

servers. These connections are made under less friendly environmental conditions and thus expected to have larger optical losses.

Let's next go through the losses from the components of the architecture for land and subsea cases. The sensing fiber in the well can be the same for the two cases, so there are no additional losses assigned to that. The umbilical fiber optic cable that leads from the IU on the FPSO to the sea floor can be from 5 km to 100 km in length. The expected optical loss from 10 km of umbilical fiber is approximately 3.4 dB. Because of this long umbilical cable, as we discussed in Section 5.3, the pulse rate will need to be reduced, which introduces an expected 3 dB loss in optical power. Because of the complicated path of the fiber from the IU to the well head, many more lossy fiber connections are required, resulting in an additional 15 dB of optical losses. Adding these losses together gives expected optical losses of approximately 21.4 dB for the subsea case over the land case.

As non-optical specialists, it is not obvious if we need to be concerned about these losses. To address that question, a synthetic DAS VSP data set has been created to mimic the land and subsea cases. Figure 5.6a shows the structural model that was used to create 71 walkaway VSP shot records. The vertical black bar shows the borehole containing the DAS fiber. Figure 5.6b shows the migrated images created for the land (dry-tree) case. Added to the 71 synthetic shot records are actual field recordings of DAS noise, which were gained appropriately to simulate the S/N noise levels for the subsea case in Table 5.1. Figure 5.6c shows the result of migrating these noisy shot records. It is clear from this example that having over 21 dB of losses from the subsea environment is significantly detrimental to the geophysical answer product. This result would not be useful for stratigraphic and time lapse studies, and it would be challenging even for structural interpretation (Wilson et al., 2021).

To mitigate these losses, four improvements to the DAS acquisition system can be made, as shown in Table 5.2 and Figure 5.7 (Ellmauthaler et al., 2020b; Willis et al., 2021). The first improvement is an upgrade to the IU to use four lasers with different wavelengths of light instead of a single laser. (A description of using lasers with

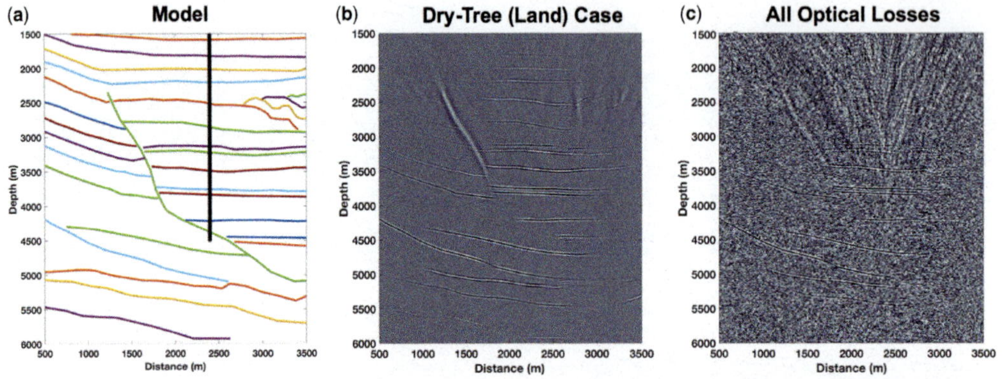

Figure 5.6. Synthetic VSP example: (a) model for the synthetics, (b) migration of land case, and (c) migration of subsea case with all losses. (From Willis et al., 2021.)

Table 5.2. Suggested improvements to the subsea DAS acquisition system (Willis et al., 2021).

Key feature	S/N gain
Quad laser DAS interrogator	2.5 dB
Active remote circulator assembly	7.0 dB
Enhanced single-mode fiber in well	10.0 dB
Raman amplification	3.0 dB
Total gain	**22.5 dB**

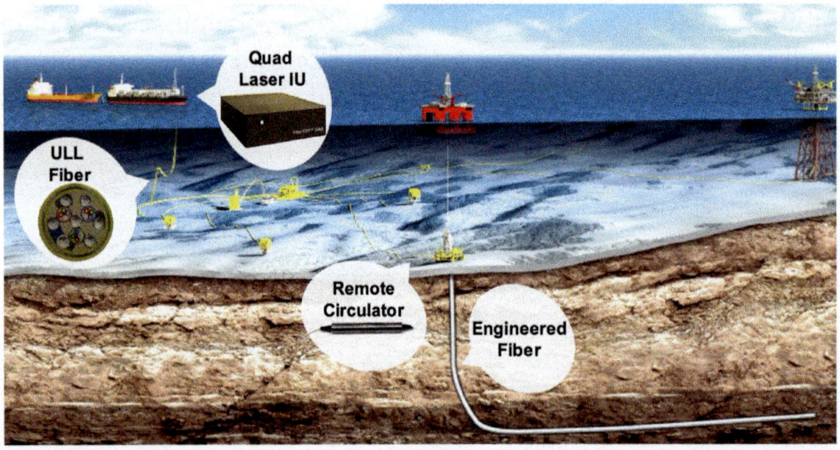

Figure 5.7. Diagram of improvements in the subsea DAS acquisition system to increase S/N. (Modified from Willis et al., 2021.)

multiple wavelengths of light will be given in Chapter 8.) This will yield an increase in S/N of 2.5 dB.

The second improvement is to add a pair of active circulators, one near the IU and one on the sea floor, as shown in Figure 5.8. A second optical fiber is added to the umbilical cable so that one optical fiber carries the pulsed light down to the sea floor and the other carries only the backscattered light up to the surface to the IU. The pair of circulators allows the down-going and up-going light to be transmitted separately. This

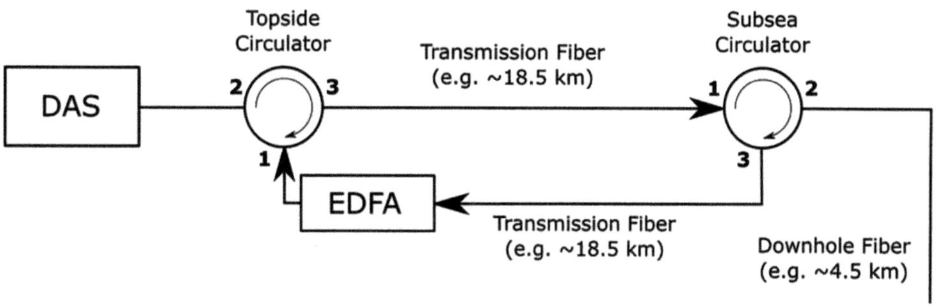

Figure 5.8. Subsea DAS topology diagram showing the addition of two circulators and the EDFA amplifier. (From Ellmauthaler et al., 2020b.)

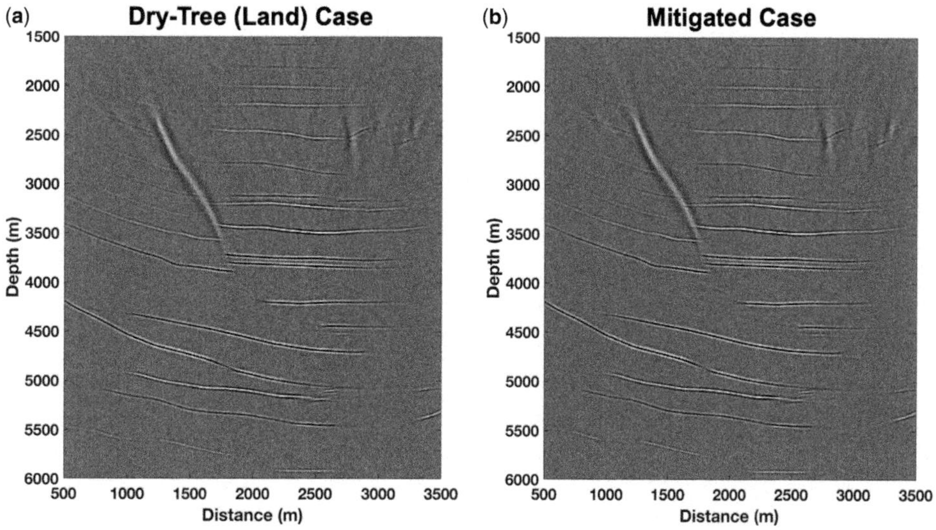

Figure 5.9. Comparison of the (a) migrated image for the land case and (b) the migrated mitigated subsea case with four improvements. (From Willis et al., 2021.)

eliminates the umbilical cable from the pulse rate computation so that only the time it takes the pulse to travel up and down the sensing fiber needs to be considered. This gives a 7 dB increase in the S/N.

The third improvement is to use enhanced backscatter (engineered) fiber for the sensing fiber in the well. This provides a 10 dB uplift in the S/N. The fourth improvement is to provide additional optical amplification of the light before entering the IU, as shown by the EDFA amplifier in Figure 5.8. This yields another 3 dB of S/N improvement.

Adding the gains in S/N for these four improvements together yields an S/N increase of 22.5 dB. This more than recovers the 21.4 dB optical losses, which means that actually this configuration is better than that in the land (dry-tree) case. Figure 5.9a shows a comparison of the migrated images for the land case alongside the mitigated case using the four improvements. The image quality of the mitigated case (Figure 5.9b) is amazingly better than the original subsea case (Figure 5.6c), and actually 0.9 dB better than the land case (Figure 5.9a).

5.5. Summary

In this chapter, I've described two different geometries for VSP DAS acquisition. The geometries of other types of surveys can vary substantially based upon the location and purpose of the survey. Land geometries tend to be less complicated and have better S/N than subsea geometries. In the VSP case, the IU, located in a recording unit, is connected to a surface or umbilical fiber optic cable which makes its way to the well head. At the well head, the fiber is connected to the sensing fiber running through the well. No matter what type of survey is acquired, accurate time information for the data is required. This is accomplished using GPS time stamping. The maximum pulse rate of the laser is limited by the length of the fiber optic cable because all of the light must propagate from the IU to the end of the sensing fiber and the back to the IU.

Because of the long umbilical required in a subsea DAS acquisition system, many factors combine to create large S/N reductions. A synthetic VSP example shows that these additional losses have a serious impact on the migrated VSP images. There are, however, several strategies available to mitigate these losses and actually more than compensate for these losses. Planning for the installation of most of these improvements needs to be done before the wells are drilled because they can't be easily retrofitted into the fiber topology.

Chapter 6 The gauge length and pulse width effects

Research is seeing what everybody else has seen and thinking what nobody else has thought.
– Albert Szent-Györgyi

6. The heart of a DAS system – the gauge and light pulse

As mentioned in Chapters 2 and 5, the gauge is what makes DAS work. It brings together light from different portions of the sensing fiber spaced apart by a distance called the gauge length. It is the heart of the optical interferometry process. In this chapter, I will explore some deeper aspects of choosing the appropriate gauge length and corresponding pulse width to determine what their effects are on the seismic data.

6.1. Following the light pulses in the fiber

It seems difficult to describe the events that happen inside in the IU and sensing fiber with just words, so let's use a series of images. Figure 6.1a is an illustration of the IU, surface cable, and sensing cable. There are two (green and blue) dots indicated on the sensing fiber separated by the same distance as the gauge length. At this moment the laser has just fired a pulse of light, indicated by the red pulse-like box, and the pulse is still inside of the IU. It then propagates through the circulator and out into the surface cable, as shown in Figure 6.1b. From there it enters the sensing fiber, placed either on the surface or in a well, as shown in Figure 6.1c.

Let's follow the pulse as it is scattered by two locations in the fiber, denoted by the green and blue dots, separated by a distance of the gauge length. First, the light is backscattered by the impurities at the green location in the sensing fiber, as shown in Figure 6.1d as the green pulse. The light pulse from the laser then continues to travel onward to the blue location and light is backscattered from there toward the IU, shown as the blue pulse. The two backscattered (green and blue) pulses then travel into the surface cable, as shown in Figure 6.1e.

Note that the distance between the two backscattered (green and blue) pulses is twice the gauge length. This is because the laser light has to travel from the green location to the blue location, be backscattered, and then travel back toward the IU. Thus, the light has to travel twice through the distance between the green and blue dots – once forward and once backward.

At this point, the two backscattered pulses enter the IU. The circulator diverts the pulses away from the path toward the laser and instead toward the direction of the receiver. Then, the light is split into two paths – one path into the gauge and the other straight toward the receiver. The gauge is actually a length of fiber equal to twice what

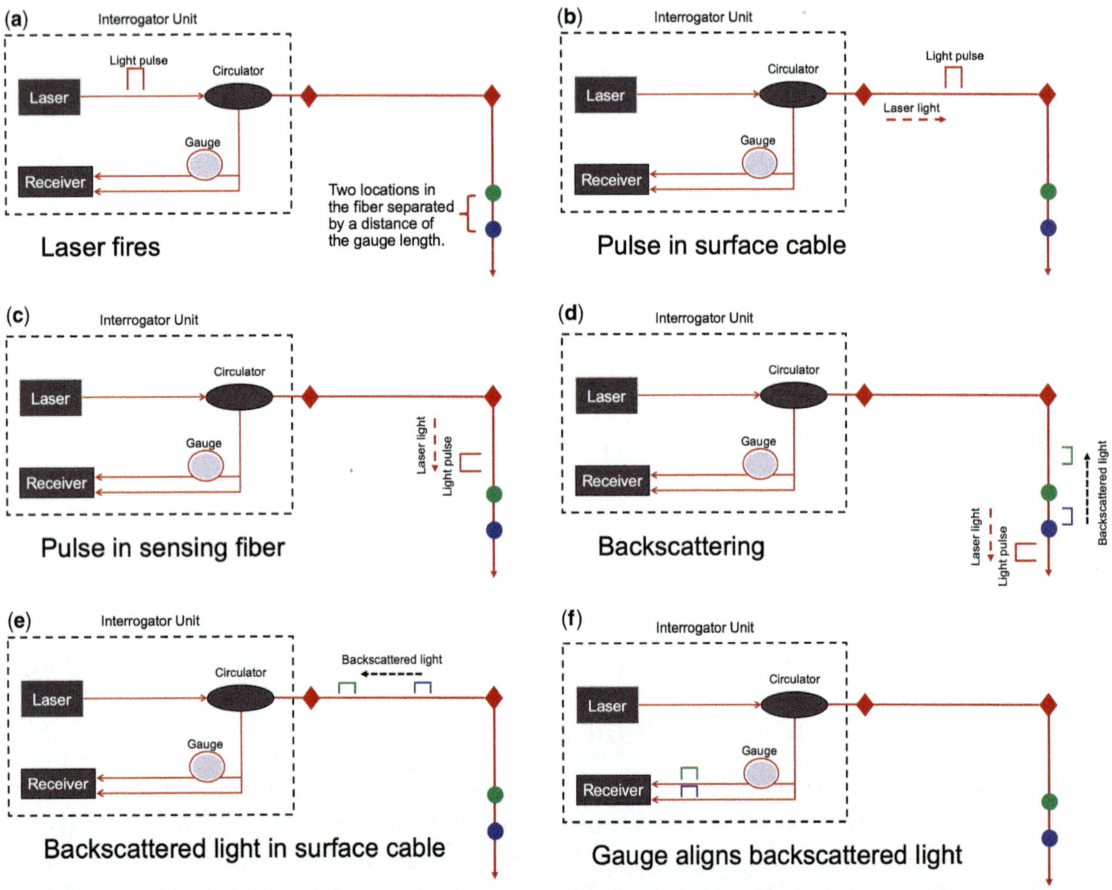

Figure 6.1. (a) The moment after the laser fires. (b) The light pulse travels inside the surface cable. (c) The light pulse reaches the sensing fiber. (d) Light backscatters off the green and blue locations on the sensing fiber. (e) The backscattered light travels in the surface cable. (f) The gauge aligns the backscattered light in the IU.

is called the "gauge length." The light that goes through the gauge is delayed by the time it takes light to propagate through the gauge. This time delay aligns the two backscattered light pulses, as shown in Figure 6.1f. Because the backscattered light from the green and blue locations are now aligned, they interfere with each other and interferometry is accomplished. The phase difference between the two pulses can be measured.

For clarity, the gauge performs a time delay of the backscattered light from the closer green point. This aligns it with the backscattered light from the farther blue point. Thus, the two-way traveltime of the light from the laser to the scattering point and back to the receiver in the IU will represent the full traveltime to the farther (blue) location.

In this discussion, I'll show the homodyne implementation to perform optical interferometry because it seems to be the easiest to understand. There are several other methods which could be used to implement this step. For example, the laser can emit two light pulses, separated by a short interval of time which simulates the aligning effect of a physical loop of fiber for the gauge (see, for example, Section 6.3.3 in Hartog [2017] or Alfataierge et al. [2020]).

6.2. Pulse width

As we have seen, the pulse of laser light from the IU is backscattered by points all along the fiber. The gauge then aligns light from backscattering points spaced apart by the gauge length. To obtain a good interferometric signal, the pulse width has to be small enough that the light from the two scattering points are completely separate in space along the fiber until they are aligned (both temporally and spatially) by the gauge in the IU.

In Figure 6.1e, the green and blue backscattered pulses are illustrated as being completely separate from each other in the surface cable. They are propagating separately all along both the sensing fiber and the surface cable. Let's view that figure and just look at the fiber cables outside of the IU, as shown in Figure 6.2a–6.2c. In Figure 6.2a, the pulse width is very short, making the two backscattered (green and blue) pulses very far apart. If the pulse width is made longer, the two pulses are closer together, as shown in Figure 6.2b. However, if the pulse is made too long, the backscattered pulses will overlap each other, as shown in Figure 6.2c, thereby degrading the interferometric process.

In Figure 6.2d, the relationship for the optimal pulse width is given as a function of gauge length for a homodyne IU (Willis et al., 2022, Section 2.4). This relationship is a function of the needed spatial gap between the backscattered pulses, the gauge length, and the speed of light in the fiber. Smaller pulse widths can be used than those

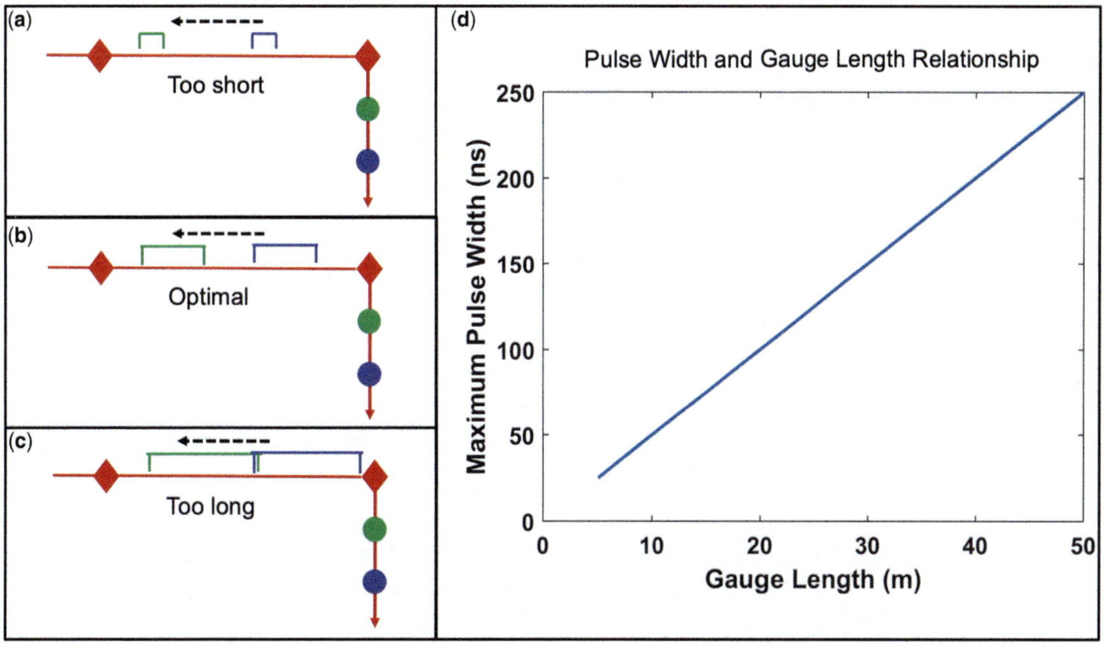

Figure 6.2. Pulse width guidelines. The left column shows three magnified views of Figure 6.1e focusing on the surface cable and sensing fiber. (a) Pulse width is too short. (b) Good pulse width. (c) Pulse width too long. (d) Relationship between the optimum pulse width and the gauge.

specified, but it will result in lower signal-to-noise ratio seismic data than would be obtained with the optimal value. Optimal light pulse shapes and widths have been studied (Dean et al., 2016b, 2018) but, for the most part, a recommended pulse shape and width will likely be provided by the IU hardware manufacturer.

6.3. The response to the gauge length

We tend to believe that a discrete sensor, such as a geophone or hydrophone, is measuring the wavefield at a single point in space. Of course, this depends on the wavelength of the acoustic energy we are attempting to measure. If we are trying to measure ultrasonic signals with an earthquake seismometer, we will be sadly disappointed because the ultrasonic wavelengths of less than 0.2 m will be very small compared to the size of the mass in the seismometer. We would need to have measurements closer than every 0.1 m to capture the ultrasonic wavefield. However, for earthquakes the dominant wavelengths are much greater than 120 m, so the seismometer is effectively making a point measurement of the wavefield.

This is also true for the measured DAS seismic signal – the wavelength of the signal to be measured matters. The DAS seismic signal is created by comparing the motion of the fiber at locations spaced apart by the gauge length. The (seismic) strain rate signal, $\dot{\epsilon}_{ZZ}^{DAS}$, that the DAS measures is given by equation 3.11 and repeated here:

$$\dot{\epsilon}_{ZZ}^{DAS} = \frac{V_z(z+GL/2) - V_z(z-GL/2)}{GL}, \tag{6.1}$$

where V_z is the particle velocity in the z-direction along the fiber, GL is the gauge length, and z is the axial direction of the fiber. If we ignore the geophone response term, from equation 3.12, this means we can simulate the DAS strain rate from the difference of the component of neighboring geophones aligned along the fiber direction, spaced apart by the gauge length using equation 6.1.

What should be obvious from this equation is that the spatial difference from the gauge length has introduced a dimension in the axial fiber direction that will affect the way the sensor will respond as a function of incident angle. This is very much like the use of arrays of geophones in surface seismic acquisition which creates a velocity filter response that attenuates slow-moving ground roll (Anstey, 1986; Morse and Hildebrandt, 1989).

For this discussion, I assume that the fiber is welded to the rock formation in the VSP case, or to the ground surface in the surface seismic case. I am not considering any coupling issues between (for example, in the VSP case) the fiber, cable, casing, cement, or formation. As shown in Chapter 3, the actual strain (rate) value of the rock is not the same as what the IU measures in the fiber.

Let's revisit the DAS strain measurement reproduced here from equation 3.2:

$$\epsilon^{DAS} = \int_{z-\frac{GL}{2}}^{z+\frac{GL}{2}} \frac{\epsilon}{GL} dz' = \epsilon \, sinc\left(\frac{k_z GL}{2}\right). \tag{6.2}$$

Again, the DAS strain rate, $\dot{\epsilon}^{DAS}$, is the simple time derivative of the strain, reproduced from equation 3.3 and expanded, which is given by

$$\dot{\epsilon}^{DAS} = \frac{d\epsilon^{DAS}}{dt} = \dot{\epsilon}\, sinc\left(\frac{k_z GL}{2}\right). \tag{6.3}$$

As mentioned previously, these relationships have been shown by multiple authors (Mateeva et al., 2014; Bakku, 2015; Daley et al., 2016; LeBlanc et al., 2018). Thus, both the measured DAS strain, ϵ^{DAS}, and strain rate, $\dot{\epsilon}^{DAS}$, are not the actual values in the fiber, but instead they are those values multiplied by an extra sinc function term, which is dependent upon the wavenumber, k_z, in the z- (axial fiber) direction and the gauge length.

We next need to convert the angular wavenumber, k_z, to something more helpful to our understanding. We know that

$$f\lambda = c, \tag{6.4}$$

$$k_z = 2\pi/\lambda, \tag{6.5}$$

$$k_z = \frac{2\pi f}{c}, \tag{6.6}$$

where f is frequency, c is apparent velocity in the axial z-direction, and λ is wavelength. Thus, we can reparametrize the sinc term in equations 6.2 and 6.3 as

$$sinc\left(\frac{k_z GL}{2}\right) = sinc\left(\frac{\pi f GL}{c}\right). \tag{6.7}$$

To dispel any confusion, let's convert the expression for the sinc term given by Dean et al. (2017), in equation 6.8, in terms of the simple wavenumber, $\tilde{k}$, to the expression (equations 6.2 and 6.10) using angular wavenumber, k_z, from Bakku (2015) (right side of equation 6.10) in the following steps:

$$sinc\left(\pi \tilde{k} GL\right), \tag{6.8}$$

$$k_z = \frac{2\pi}{\lambda} = 2\pi\tilde{k}, \text{ or } \tilde{k} = \frac{k_z}{2\pi}, \tag{6.9}$$

$$sinc\left(\pi \tilde{k} GL\right) = sinc\left(\pi \frac{k_z}{2\pi} GL\right) = sinc\left(\frac{k_z GL}{2}\right). \tag{6.10}$$

Be aware that, in signal processing, the usual definition of the sinc function (Bracewell, 1965, page 62) incorporates the factor of π, which gives the following definition:

$$SINC(x) = \frac{\sin(\pi x)}{\pi x} = sinc(\pi x), \tag{6.11}$$

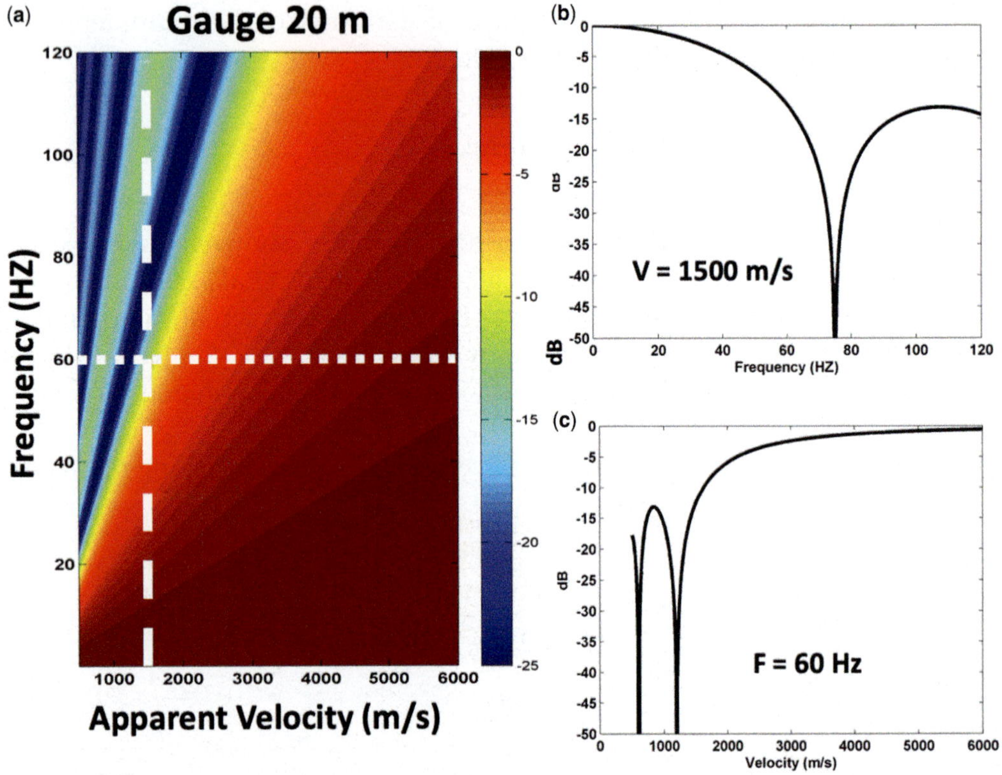

Figure 6.3. Effect of a 20 m gauge. (a) Spectral pass band of the gauge. (b) A cut through the spectrum at $v = 1500$ m/s. (c) A cut through the spectrum at $f = 60$ Hz. (From Willis et al., 2014.)

where I am using *SINC* to indicate the signal processing definition of the function. This can lead to confusion when using libraries of functions to evaluate the sinc function.

It is important to understand the effect this sinc function has upon the seismic strain (rate) data that are acquired. Willis et al. (2014) presents graphs showing the spectral response of the sinc term for a 20 m gauge as a function of apparent velocity (on the horizontal axis) and frequency (on the vertical axis), as shown in Figure 6.3a. The color indicates the filter response in dB that the sinc imparts to the strain (rate) data. To clearly illustrate this, there are two white lines through the figure. The dashed line, at v = 1500 m/s, is a cut through the response surface which is plotted as a curve in Figure 6.3b. There is a big notch (zero) in the spectrum at about 75 Hz with steeply sloping amplitudes on either side. This means the measured strain (rate) for a wave propagating at 1500 m/s (water velocity) would be significantly attenuated above 60 Hz. A cut through at a constant frequency of 60 Hz (white dots in Figure 6.3a) is shown in Figure 6.3c. This shows that the strain (rate) signal which we measure for any wave that propagates with an apparent velocity less than approximately 1500 m/s will be significantly attenuated.

Figure 6.4 shows the spectral response for four different gauge lengths: 5 m, 10 m, 20 m, and 40 m. We must make sure that the data we need to record fall into the frequency and apparent velocity range far away from the notches in the spectrum. It is clear that the smallest gauge of 5 m gives the largest pass band while the largest gauge of 40 m gives the most restricted pass band range of frequencies and apparent velocities.

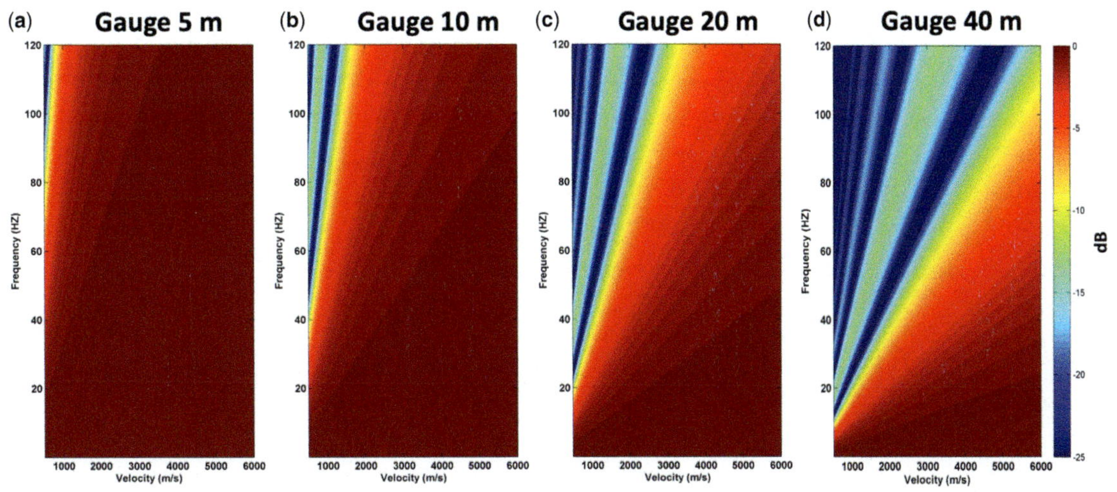

Figure 6.4. Spectral response of the sinc function for gauges of size (a) 5 m, (b) 10 m, (c) 20 m, and (d) 40 m. (From Willis et al., 2014.)

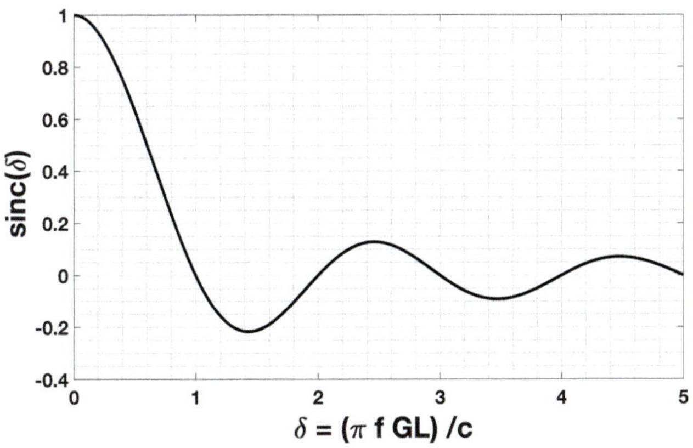

Figure 6.5. Sinc function showing the dimensionless scaling factor, δ, as the horizontal axis.

Spectral plots of the sinc function term have been shown and discussed by many other authors (Dean et al., 2015a, 2015b, 2019; Bona et al., 2017; Karrenbach et al., 2019; Mizuno et al., 2019; Jestin et al., 2020; Kobayashi et al., 2020) to help understand the tradeoffs between large and small gauge lengths. It is known that the signal-to-noise ratio increases as the gauge length is increased (Dean et al., 2015b; Willis et al., 2022). This means that while smaller gauges give better spectral performance, longer gauges provide higher signal-to-noise ratio data sets.

From equation 6.7, we can build a dimensionless plot of the sinc function which covers all cases of frequency and apparent velocity. My college theoretical physics and mechanics instructors have always loved these types of plots, but I have always struggled to make use of them. I will include Figure 6.5 as an example in case you find it helpful. Note the notches (zeros) occur when the normalized parameter, δ, is a nonzero integer. While not obvious on Figures 6.3a and 6.4, at the location of each notch the response flips sign.

6.4. Examples of the effect of the gauge

Looking at the response curves may not be the most intuitive way to understand the effect of the gauge on the strain (rate) data. Let's look at some actual wavelets. In Figure 6.6, I've plotted the strain rate wavelets for different center frequency Ricker

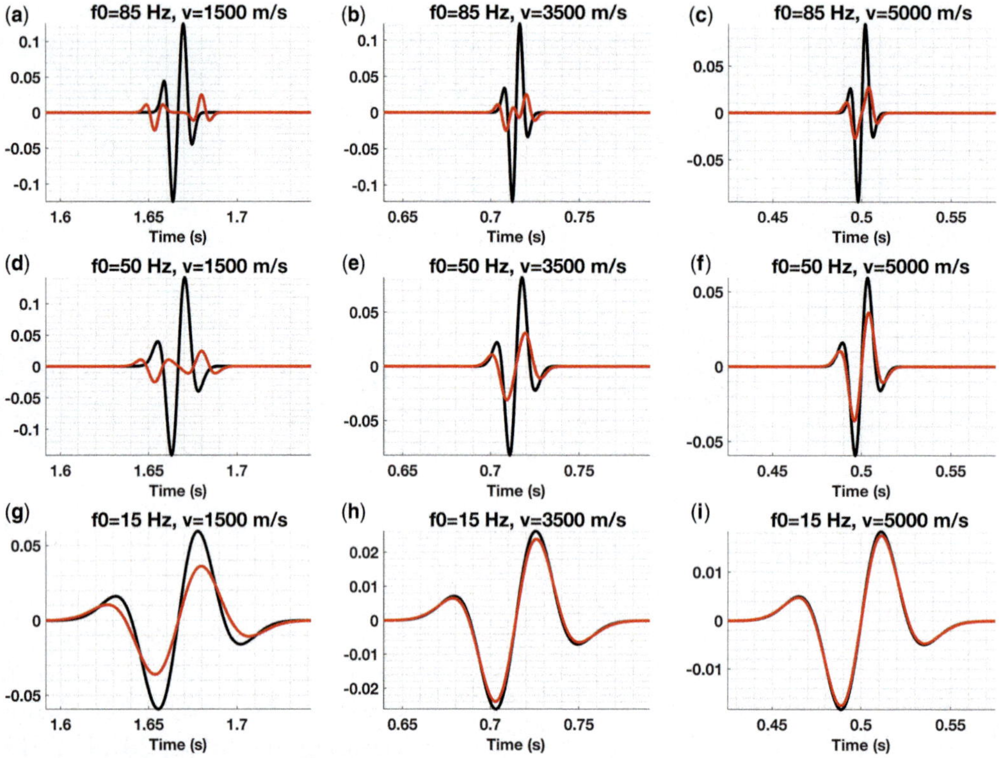

Figure 6.6. Strain rate traces for different source frequencies, apparent velocities, and gauge lengths. Results for the 10 m and 40 m gauges are shown in black and red, respectively.

wavelets, different apparent velocities, and different gauge lengths. I've used equation 6.1 to generate these wavelets by first creating the V_z traces with Ricker wavelets and then differencing them. That is why the strain rate wavelets don't look like pure Ricker wavelets.

The left, center, and right columns in Figure 6.6 have apparent velocities of 1500 m/s, 3500 m/s, and 5000 m/s, respectively. The top, center, and bottom rows have source center frequencies of 85 Hz, 50 Hz, and 15 Hz, respectively. The 10 m and 40 m gauge traces are in black and red, respectively.

Let's see what information we can extract from these images.

1. The $f_0 = 15$ Hz source (bottom row) is usable for both gauge lengths and all apparent velocities.
2. The $f_0 = 50$ Hz source (middle row) fails for the slow 1500 m/s velocity with the larger 40 m gauge (red), but it is reasonable for the other apparent velocities and the smaller gauge.
3. The $f_0 = 85$ Hz source (top row) is reasonable only for the fast 5000 m/s apparent velocity for the larger 40 m gauge (red), and it is reasonable for all but the slow 1500 m/s apparent velocity for the 10 m gauge.
4. A larger gauge appears to lower the overall amplitude of the strain rate measurement.

It is clear now from these examples that the appropriate gauge should be selected for the desired application. Earthquake monitoring applications can use very large gauges without any significant loss in signal bandwidth because they have very low-frequency content. Deep water Gulf of Mexico VSP data sets, where the maximum expected frequency in the data is likely to be only 20 Hz, also can use larger gauges. However, applications where it is desired to monitor shallow slow-velocity formations, or S-wave surveys, it is very likely that small gauge lengths will be required to preserve the bandwidth of the seismic signals.

Several authors have developed expressions to select the optimal gauge length. Equation 3 from Dean et al. (2015b) seems to recommend the maximum gauge length to be

$$GL = \frac{2V_{min}}{3f_{dom}}, \tag{6.12}$$

where V_{min} is the slowest velocity of interest and f_{dom} is the dominant frequency of the data. The equation in Cuny et al. (2020) modifies this recommendation to allow for 1 dB maximum spectral distortion to give the following best value for the gauge length:

$$GL \approx 0.47 \cdot \frac{V_{apparent}}{f_{max}}, \tag{6.13}$$

where $V_{apparent}$ is the apparent velocity of the slowest event of interest and f_{max} is the highest frequency of interest. The optimal gauge length from equation 6.13 is shown graphically in Figure 6.7. For each of the curves of constant gauge length size, it is safe to use that value for all frequencies below the curve and all apparent velocities to the right of it. From this, it is clear that a 10 m gauge is good for most exploration seismic

Figure 6.7. Optimal gauge lengths given by equation 6.12. Colors indicate the recommended maximum gauge length for the desired maximum frequency (vertical axis) and apparent velocity (horizontal axis).

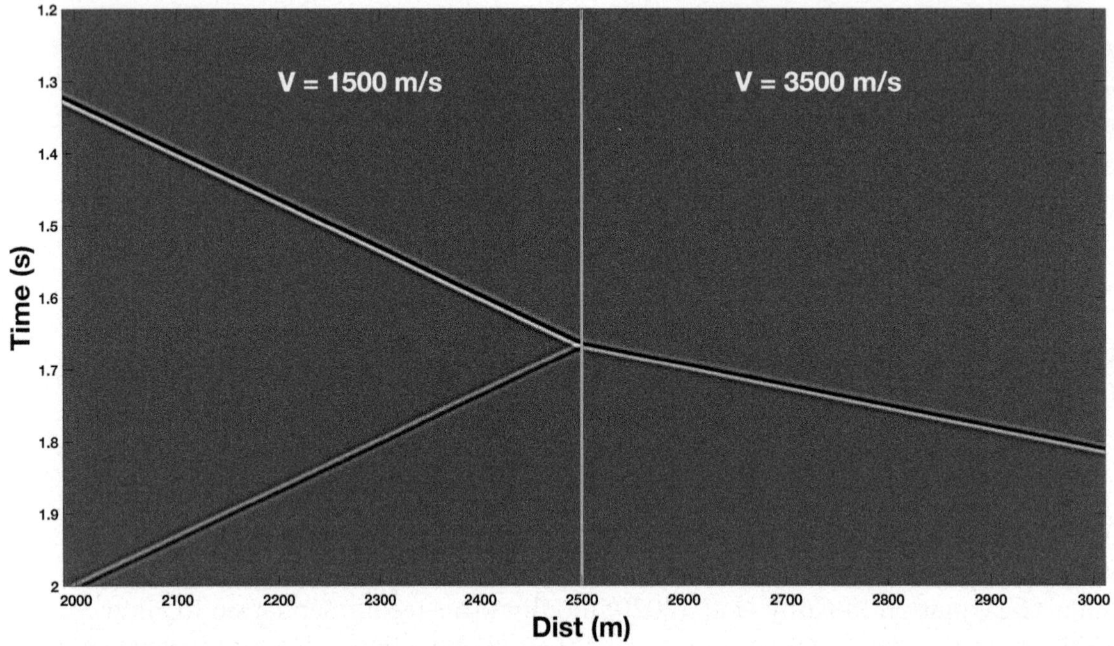

Figure 6.8. Simple two-layer synthetic strain rate VSP record.

projects where the maximum frequency band is 90 Hz and the slowest apparent velocity is 2000 m/s.

6.5. *Viewing the effect of the gauge in the f-k domain*

In many ways, the easiest way to visualize the effect of the gauge is in the frequency-wavenumber (*f-k*) domain. Figure 6.8 shows a synthetic VSP strain rate record with two layers. The upper layer is (unbelievably!) 1500 m/s and the lower layer is 3500 m/s. These values have been chosen to pick some of the velocity extremes to illustrate the gauge. The source center frequency of the Ricker wavelet is 70 Hz.

The top row of Figure 6.9 shows the (absolute value of the) sinc function for two different gauges. The bottom row shows the *f-k* spectrum for the strain rate data using these same two gauge lengths. I've included yellow tick marks where the sinc function has zero values.

Figure 6.10 shows the *f-k* spectra for four different gauges: a) 5 m, b) 10 m, c) 20 m, and d) 40 m.

6.6. *Changing the gauge length after acquisition*

The role of the gauge is to force together backscattered light from two different locations on the fiber so that their phase angle can be determined. Having this performed by hardware is efficient because the phase angle extraction can be performed in real time and the signal-to-noise ratio can be at its highest due to the additional processing and optimization that can be performed. The signal quality is optimal because the pulse width is chosen specifically for the selected gauge length, using the most amount of light that is practical. It is also optimal due to the full fidelity of the

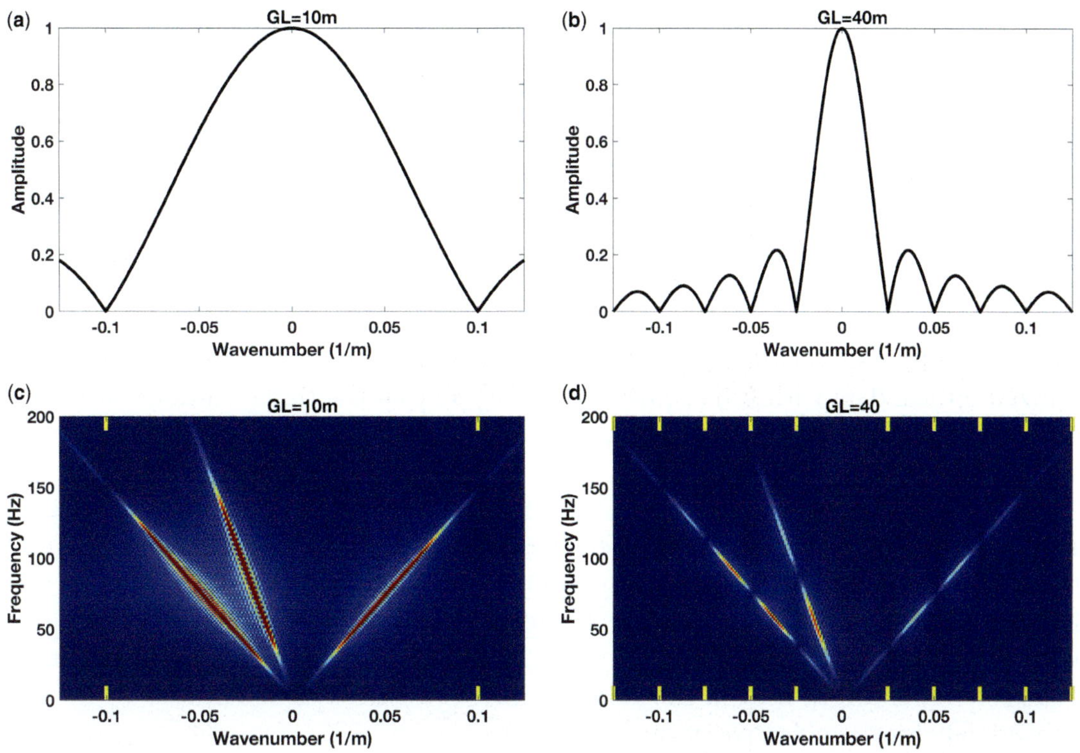

Figure 6.9. The effect of the two different gauges shown by their sinc functions (top row) and the corresponding *f-k* spectra (bottom row).

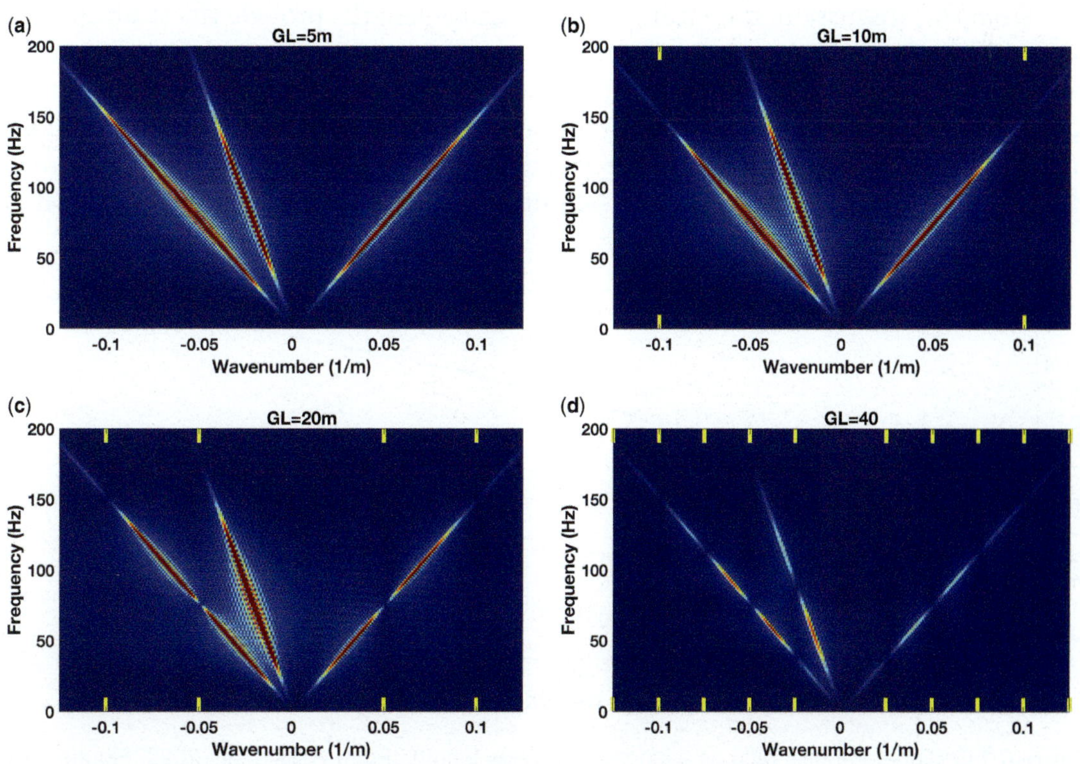

Figure 6.10. The *f-k* spectra of synthetic strain rate data acquired with gauges of (a) 5 m, (b) 10 m, (c) 20 m, and (d) 40 m.

optical interferometry data available in real time which is processed in the field and enhanced using channel stacking and frequency down sampling.

An alternative approach is merely to record the backscattered light stream directly, without performing the optical interferometry step in hardware, and then later post-process the signal, implementing the interferometry step digitally when back in the office (Cuny et al., 2020). The value of this approach is the ability to select any one (or series of) arbitrary gauge length(s) after acquisition and to combine them in a depth-variable manner. A limitation is the requirement to select the smallest pulse width for any anticipated gauge length that will be used, making the pulse width suboptimal for large gauge lengths. A challenge most likely will be the requirement to record the digitized optical data with full acquisition fidelity so that the signal-to-noise ratio won't suffer if the data needs to be subsampled in the field to make it practical to store and transfer back from the field.

Another approach is to record the field data using two or more different (optical) gauge lengths as different data sets. Then these data sets can be recombined back in the office to create optimized seismic strain (rate) data (Wu et al., 2019). This method assures that the best pulse width is selected for each data set and that the optimal processing of the optical data is performed. The limitation is that only a finite number of gauge lengths can be used in the field, so the lengths need to be selected appropriately for the objective of the survey.

6.7. Effect of gauge length on signal-to-noise ratio

From this discussion, it is clear that small gauge lengths provide the strain (rate) data with the least distortions. So, it seems obvious that we want to use a very small gauge. The downside of a small gauge, however, is that the signal-to-noise ratio of the data is bigger than with a large gauge. Willis et al. (2022) claim that the signal-to-noise ratio increases as the square root of the gauge length, as shown in Figure 6.11. Figure 6.12 shows an example of a VSP record acquired with two different gauge lengths. It is

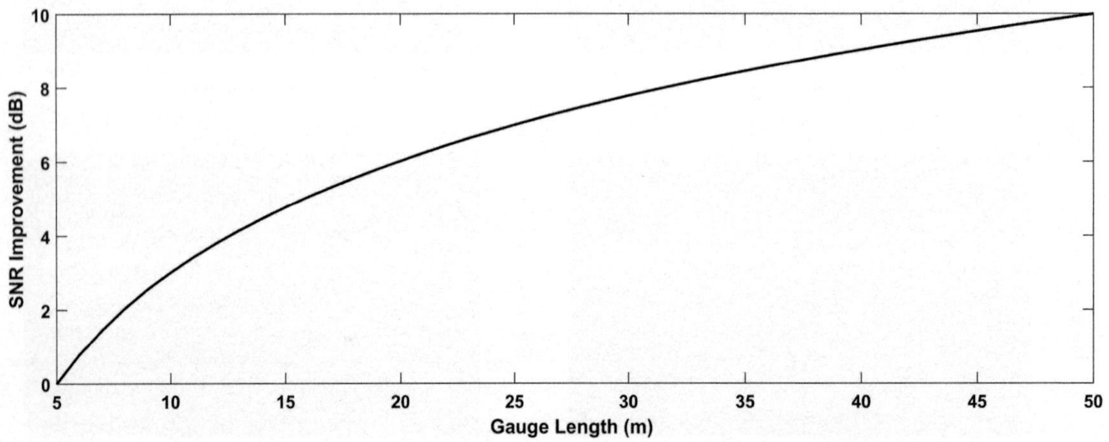

Figure 6.11. Signal-to-noise ratio as a function of gauge length. (From Willis et al., 2022, Figure 2.3.)

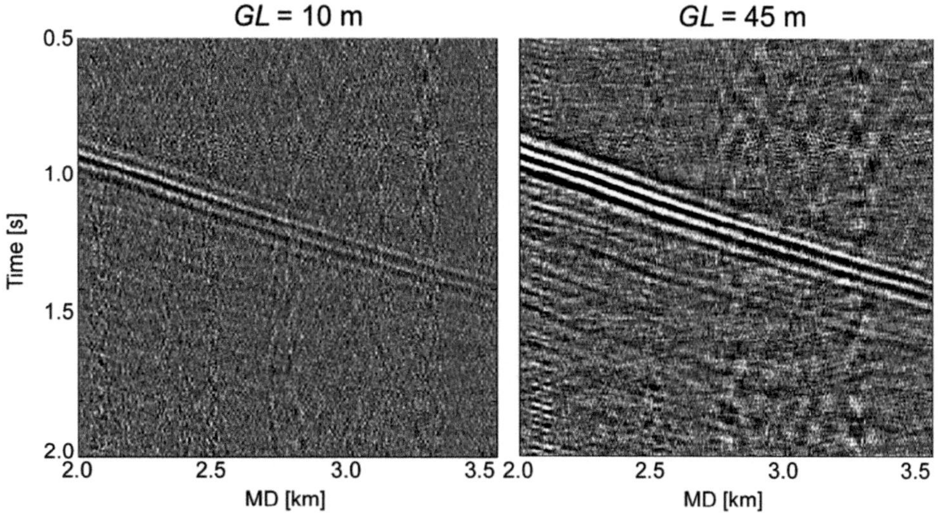

Figure 6.12. Examples of VSP records recorded with two different gauges. (Modified from Kobayashi et al., 2020, Figure 5e.)

clear that the 10 m gauge length record has a higher noise level compared to the signal than the record using a 45 m gauge.

6.8. *Summary*

The gauge is at the heart of the DAS interferometry process. For simplicity, I've described the homodyne implementation of optical interferometry. A laser in the IU fires individual pulses of light into the surface cable which is connected to the sensing fiber either on the ground or in a borehole, for example. The duration of the pulse of light from the laser, called the pulse width, most likely will be predetermined by the vendor of the IU for each gauge length, so it won't need to be selected as an independent acquisition parameter.

Variations in the sensing fiber cause the light pulse to be backscattered toward the IU. Inside the IU, the backscattered light is guided to a direct path and also to a delayed path en route to the receiver. The delayed path contains a gauge which is simply a loop of fiber with a length equal to twice what is called the gauge length. The optical interference of the light from the two paths allows the computation of the phase difference between points all along the sensing fiber separated by the distance of the gauge length. Then, the computed phase angle is converted to strain (rate) for channels of data spaced apart by about 1 m all along the sensing fiber.

The selection of the gauge length is important because it dictates the range of apparent velocities and temporal frequencies that will be recorded without distortion. Large gauge lengths will improve the signal-to-noise ratio of the data, while small gauges will preserve a larger range of apparent velocities and frequencies. So, the selection of the gauge length should be determined by the objectives of the survey – specifically the required frequency content of the signal and the apparent velocities of the seismic waves.

Chapter 7 Angle of incidence

Your argument is sound, nothing but sound.

– Albert Einstein

If you don't believe in logic, you might as well give up science.

– Keiiti Aki

7. Angle of incidence response of fibers

A geophone is designed to measure a single component of the seismic wavefield. Combining one vertical and two horizontal geophones allows the entire seismic wavefield to be measured. In a similar manner, an optical fiber can be used to measure a single component of the wavefield. Some clever modifications of the fiber optic cable and/or layout of the cable can expand the response of the fiber optic cable to measure more than just one component. In this chapter, I will briefly review the response of a geophone and then go into some detail about the angular response of fiber.

7.1. Angular response of a geophone

From Chapter 3 of Bakku (2015), the vertical component of particle velocity, v_z, from an incident P-wave upon a geophone can be written as

$$v_Z = A_0 \cos\theta e^{i(kz-\omega t)}, \tag{7.1}$$

where A_0 is the scalar amplitude of the source, θ is the angle of incidence of the wave to the vertical axis of the geophone, $K = \omega / \alpha$, ω is angular frequency, α is the P-wave velocity, and t is time.

The vertical component of particle velocity, v_z, from an incident S-wave upon a geophone is written as

$$v_Z = A_0 \sin\theta e^{i(kz-\omega t)}. \tag{7.2}$$

Choosing $A_0 = 1$ allows us to create a polar plot of the P- and S-wave responses, shown in Figure 7.1a and 7.1b, respectively. This shows a perfect sensitivity when P-waves propagate at 0 and 180 degrees, and when S-waves propagate at 90 and 270 degrees. The solid lines are positive amplitudes, while the dashed lines are negative values.

I find that the polar plots give a great overview of the sensitivity curves but the polarity always seems confusing. To better illustrate the polarities as a function of incident angle, I've replotted the same curves on a linear axis in Figure 7.2.

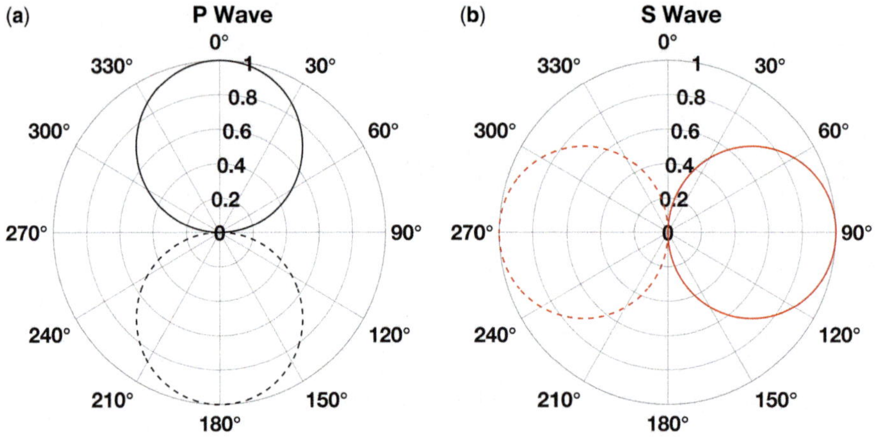

Figure 7.1. Normalized geophone V_z sensitivity for (a) P-waves and (b) S-waves.

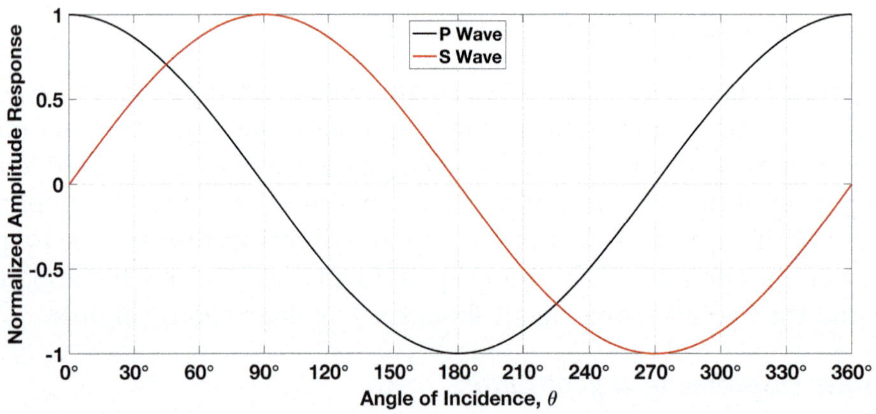

Figure 7.2. Normalized geophone V_z sensitivity to an incident P-wave (black) and S-wave (red).

7.2. Angular response of fiber

For this discussion, I will omit the effect of the gauge, or, in other words, assume the gauge is infinitely small. Using equation 3.4 from Bakku (2015), the vertical component of strain, ϵ_{zz}, from an incident P-wave upon a fiber can be written as

$$\epsilon_{ZZ} = \frac{-A_0}{\alpha} \cos^2\theta \; e^{i(kz-\omega t)}, \tag{7.3}$$

where α is the P-wave velocity.

Now using equation 3.7 from Bakku (2015), the vertical component of strain, ϵ_{zz}, from an incident S-wave upon a fiber is written as

$$\epsilon_{ZZ} = \frac{-A_0}{2\beta} \sin 2\theta \; e^{i(kz-\omega t)}, \tag{7.4}$$

where β is the S-wave velocity.

For simplicity, we will choose $\frac{\alpha}{\beta} = 1.5$. This time I will normalize the strain responses by a factor of $\frac{-A_0}{\alpha}$, which allows us to create a polar plot of the P- and S-wave strain responses, shown in Figure 7.3a and 7.3b, respectively. This shows perfect sensitivity when P-waves are propagating at 0 and 180 degrees. However, for S-waves, the sensitivity is much less and more complex with four lobes of alternating sign. This is again perhaps better displayed on a linear axis, as shown in Figure 7.4.

Let's put all of these together to understand how they are connected. Figure 7.5 shows V_z in dotted lines and strain (ϵ_{zz}) in solid lines. The P-waves are in black lines, while S-waves are in red lines.

We see that the P-wave strain sensitivity is always positive, while the geophone flips polarity with the direction of the wave propagation. We notice that the normalized P-wave strain response is identical to the normalized magnitude of the geophone response at 0°, 90°, 180°, and 270°. The greatest losses for the P-wave strain compared to the normalized magnitude of the geophone occur at 45°, 135°, 225°, and 315°.

Because the S-wave sensitivity between the geophone and strain are so different, it is difficult to make easy comparisons. The biggest difference between them is that, in

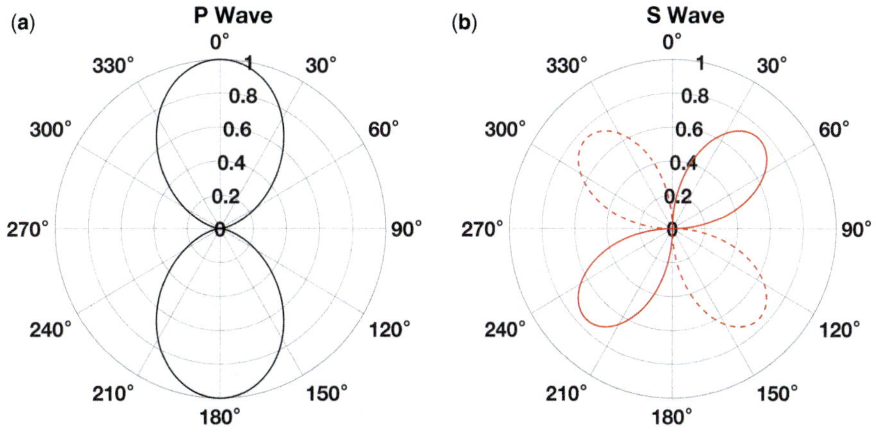

Figure 7.3. Normalized strain sensitivity for (a) P-waves and (b) S-waves.

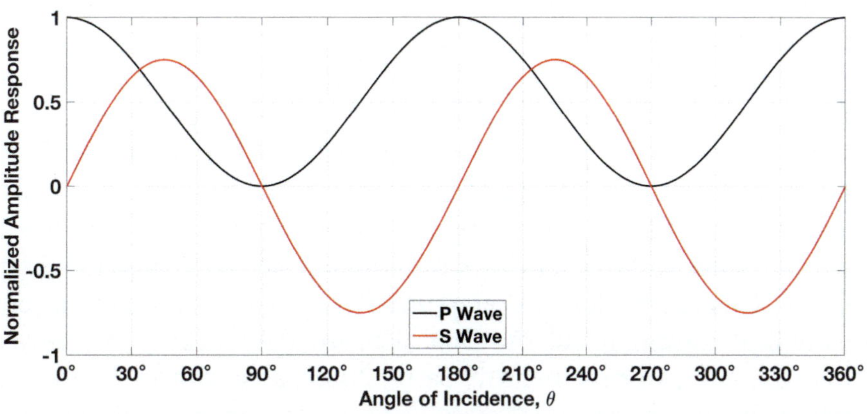

Figure 7.4. Normalized sensitivity of strain to an incident P-wave (black) and S-wave (red).

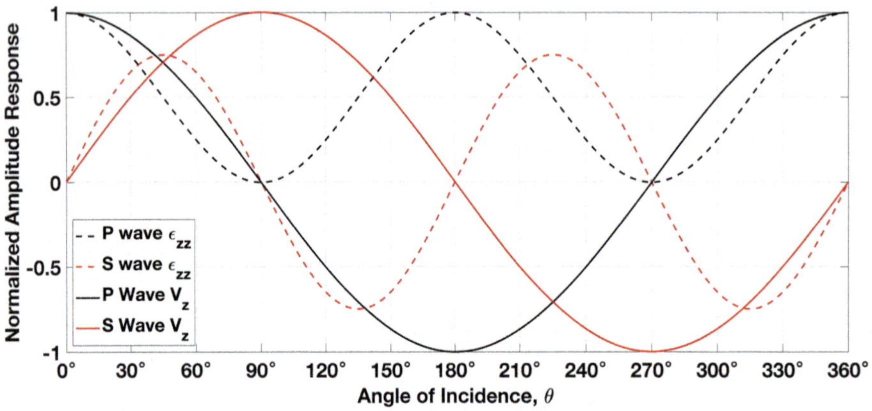

Figure 7.5. Normalized sensitivity of a geophone V_z (solid lines) and strain (dotted lines) to an incident P-wave (black) and S-wave (red).

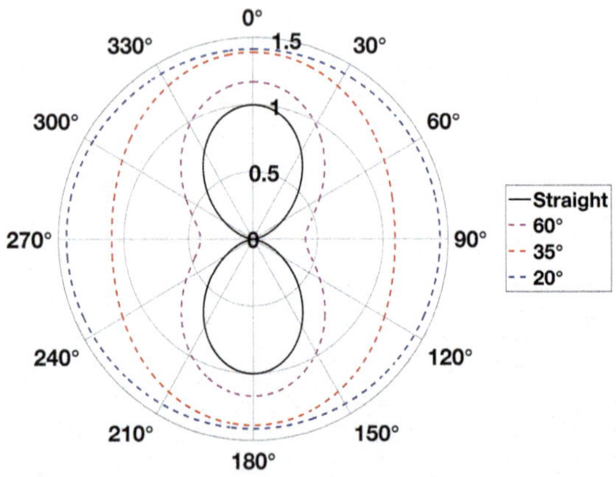

Figure 7.6. Angular P-wave response for a straight fiber (black), and helical wound fibers with winding angles of 60° (magenta), 35° (red), and 20° (blue).

360°, the geophone response goes through one complete cycle while the strain response goes through two complete cycles.

7.3. Angular response of helically wound fiber

The fiber optic cable can be used to measure strain all along the fiber as I've described in Section 2.2. One important aspect is that the extracted strain measurement is a single component of the 9C strain tensor in the axial, or length-wise, direction of the fiber. As such, it has a sensitivity profile for P-waves, shown by the black line in Figure 7.3 and repeated in Figure 7.6, which has been documented by many authors (Lumens et al., 2013; Mateeva et al., 2014; Bakku, 2015; Willis et al., 2016c). The axial direction of the fiber is 0°, which shows full sensitivity of 1 for P-waves traveling in the axial direction. The sensitivity of the fiber to strain changes as the angle of incidence of the P-wave is increased, making a dumb bell-like shape. It is clear from this response curve that the fiber is completely insensitive to P-waves traveling normal (90° and 270° in the figure) to the axial direction (0° and 180°) of the fiber.

In many applications, it is important to be able to measure the seismic wavefield in more than only a single direction. One approach is to attempt to modify the path of the optical fiber in the cable to make its response more like an omnidirectional sensor, such as a hydrophone. This can be done by wrapping the optical fiber around a center mandrill in a helical manner inside the cable (Hornman et al., 2013; Mateeva et al., 2014; Poletto et al., 2015; Hornman, 2017; Baird, 2020), as shown in Figure 7.7. The winding pitch of the fiber can be selected to optimize the desired response. I define

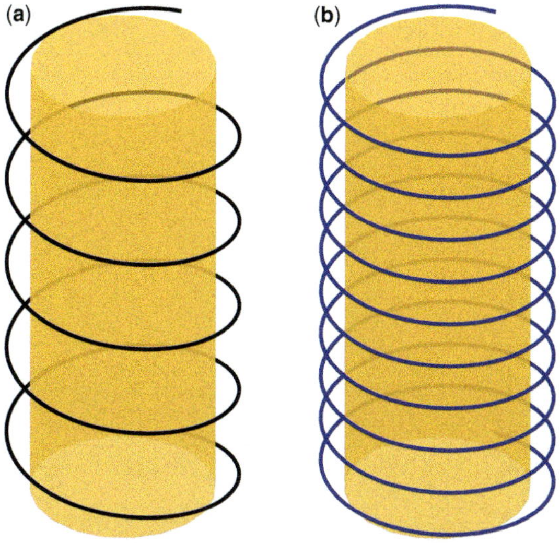

Figure 7.7. (a) Schematic of a helically wrapped fiber optic cable. The fiber is in black, the mandrill is orange. (b) Schematic of a helically wrapped fiber optic cable with half the pitch angle, which doubles the amount of fiber per unit distance of cable. The fiber is blue.

Figure 7.8. Comparison of three types of fiber and a geophone. The black cable is an armored single strand fiber, the white cable is a single strand of fiber, the blue cable is a helically wrapped fiber cable, and at the right is a geophone. (From Spikes et al., 2019, Figure 1.)

a winding pitch of 90° as the "degenerate" straight fiber case, 45° as the case of the fiber moving equally around the mandrill and up the mandrill, and 0° as simply wrapping around the mandrill without moving up along the cable.

Kuvshinov (2016) derive the P-wave sensitivity of a helically wound fiber perfectly coupled to the wellbore. Using Kuvshinov's equation 20 for a perfectly coupled fiber, and the elastic parameters for the fiber scenario used for Kuvshinov's Figure 3, I've created the normalized angular response curves for three different winding angles, which are shown in the dotted lines in Figure 7.6. It is clear that the 60° winding angle (in magenta) is only a very small improvement in sensitivity for directions other than 0° and 90°. However, for 35° (in red) and 20° (in blue) winding angles, there is significant improvement, providing an almost full range of angle sensitivity.

Given this great improvement in angular sensitivity, it seems obvious that helically wound fiber is the solution we seek. However, in order to preserve the waveguide geometry inside the fiber, care must be taken so that the radius of curvature of the fiber is not made too small. This requirement forces the cable to be larger in diameter than would be needed for a single straight fiber, as can be seen by comparing the black single-strand fiber cable with the blue helically wound fiber cable shown in Figure 7.8.

The drawback of using helically wrapped fibers is that S-wave sensitivity is greatly reduced, especially for wrapping angles of 35° (Baird, 2020). Figure 7.9a and 7.9b shows synthetics for what would be recorded for a microseismic event with a straight fiber and a helically wrapped fiber with a wrapping angle of 35°, respectively. Note that the apex of the P-wave has better sensitivity on helically wrapped fiber, but the S-wave has completely disappeared.

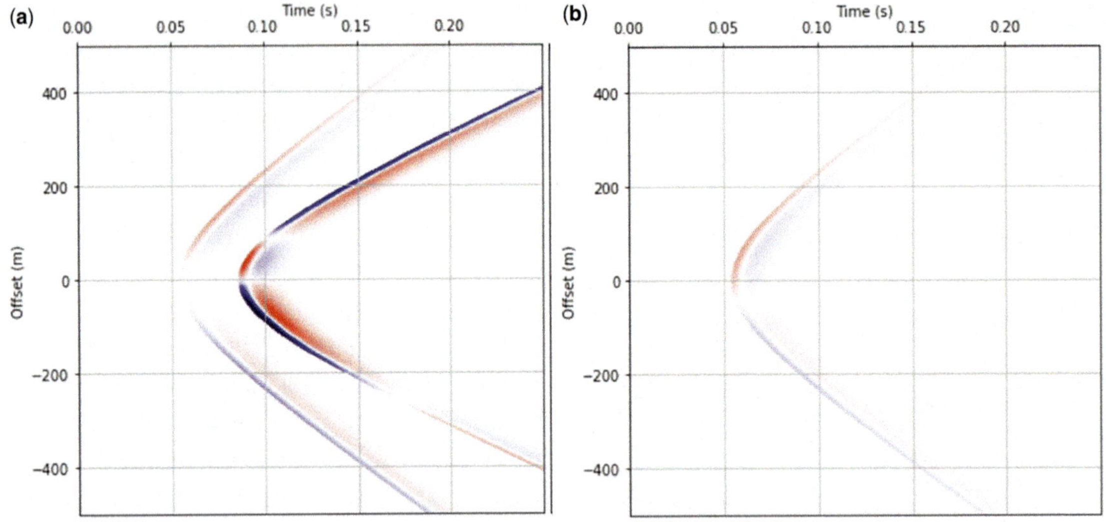

Figure 7.9. Synthetic microseismic event that would be recorded by (a) a straight fiber and (b) a helically wrapped fiber at 35°. (Modified from Baird, 2020, Figure 2.)

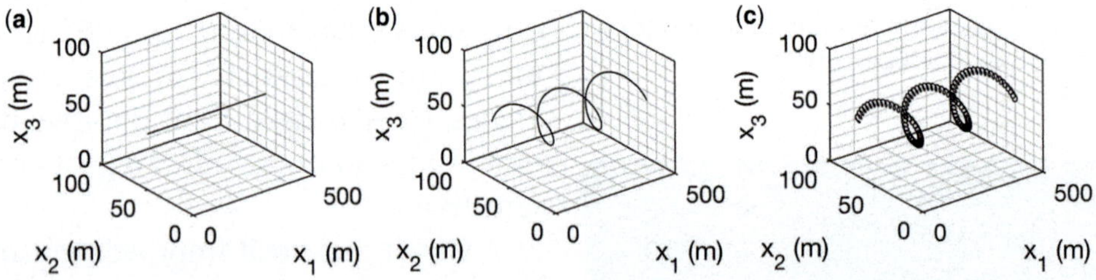

Figure 7.10. Nomenclature for multiple helices: (a) 0-helix, (b) 1-helix, and (c) 2-helix. (From Innanen, 2017a, Figure 2.)

7.4. Increasing the sensitivity of helically wrapped fibers

Innanen (2017a) defines terminology for the nesting of helical spirals in a single fiber as n-helices, where n is the order of twisting, as shown in Figure 7.10. A straight fiber has an order of n = 0, a single helix fiber has an order of n = 1, and a double compound helix has an order of n = 2. Figure 7.11a and 7.11b compares the synthetic responses of the 0-helix to the 2-helix, respectively, for a simulated explosive source. The 0-helix (i.e., straight) fiber has clear broadside insensitivity at a 250 m arc length along the fiber. The 2-helix (double helix) fiber has a rapidly changing regular pattern of low- and high-sensitivity zones and spans a much longer arch length because the fiber is spiraling through a much longer path but covering the same axial distance. Thus, the 2-helix fiber has improved sensitivity at 700 m which corresponds to the insensitive portion of the 0-helix at 250 m. However, additional processing is required to remove the high/low pattern in the strain.

Several authors have studied the optimal cable designs and responses for this architecture (Innanen 2017a, 2017b; Eaid et al., 2018; Wilks and Wuestefeld, 2018; Wuestefeld and Wilks, 2019) and even created corkscrew well paths to obtain similar helical behavior (Wuestefeld, 2021).

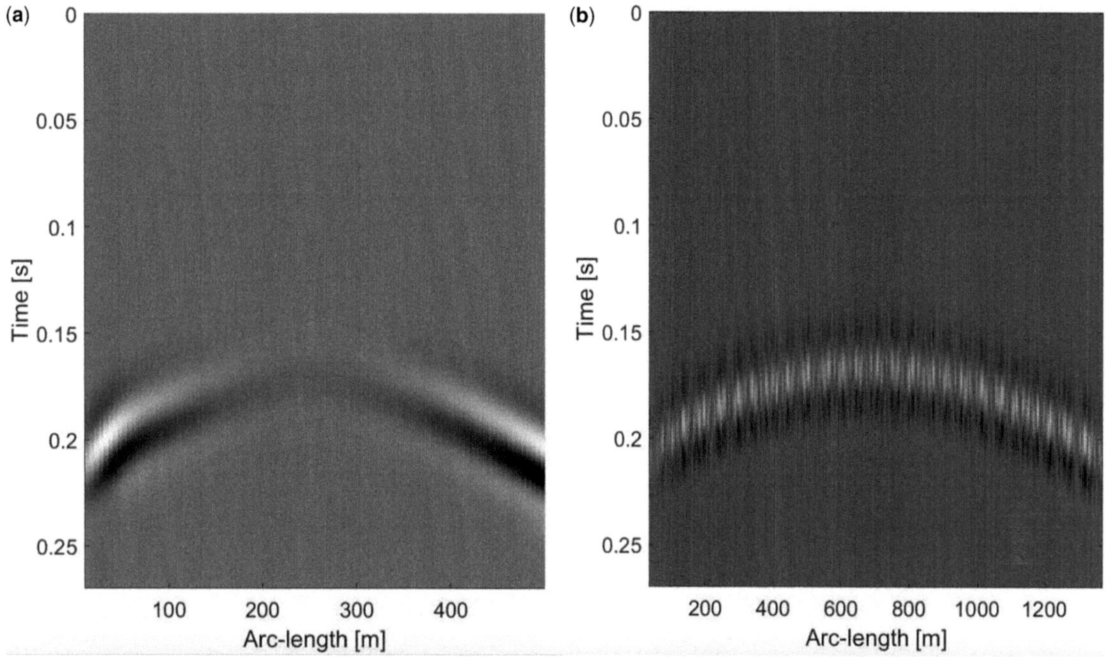

Figure 7.11. Comparison of synthetic explosion using (a) straight fiber and (b) 2-helix fiber. (Modified from Eaid et al., 2018, Figures 5 and 7.)

The challenge with this method is that it requires small gauge lengths and the wavelength of the seismic signal must be much larger than the helical dimensions (Eaid et al., 2020a). An additional challenge is that with each higher order of helix will come the need for a larger form factor which may make it impractical in applications where space is critical, such as in wellbores.

7.5. Creating three-component DAS surveys using orientable sources

We have seen that the optical fiber is a single component measurement. By changing the geometry of the fiber, using helical wrapping around a mandrill, its effective response can be modified to be more omnidirectional. For surface seismic exploration applications, this omnidirectional response may be ideal. However, in other applications, including earthquake and microseismic monitoring, three components (3C) of the seismic wavefield are strongly preferred.

By invoking reciprocity, Erdemir et al. (2019) teach using directional active sources (e.g., vertical and shear wave vibrators, or an orientable weight drop source) to create three component seismic measurements with DAS. Figure 7.12 shows a schematic of the process for a VSP application. While the DAS fiber still measures the vertical component of motion, by recording the DAS fiber response separately for each of the three source orientations, the three components of the seismic wavefield can be acquired and separated. Tertyshnikov and Pevzner (2020) demonstrate this method using an orientable weight drop source, as shown in Figure 7.13. The separation of the P-waves onto the Z component and the S-waves onto the X and Y components show that it is indeed quite effective.

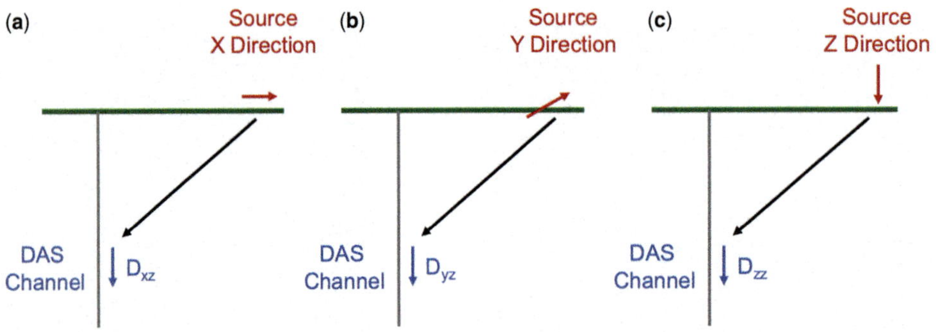

Figure 7.12. Schematic showing how to create 3C DAS data with a directional source. The red arrow is the source, the black arrow is the seismic ray path, and the blue arrow is the direction of measured DAS strain data. (a) The X component created with shear wave source in the X direction, (b) the out of plane Y component created with a shear wave source in the Y direction, and (c) the vertical component created with a vertical source. (Modified from Erdemir et al., 2019.)

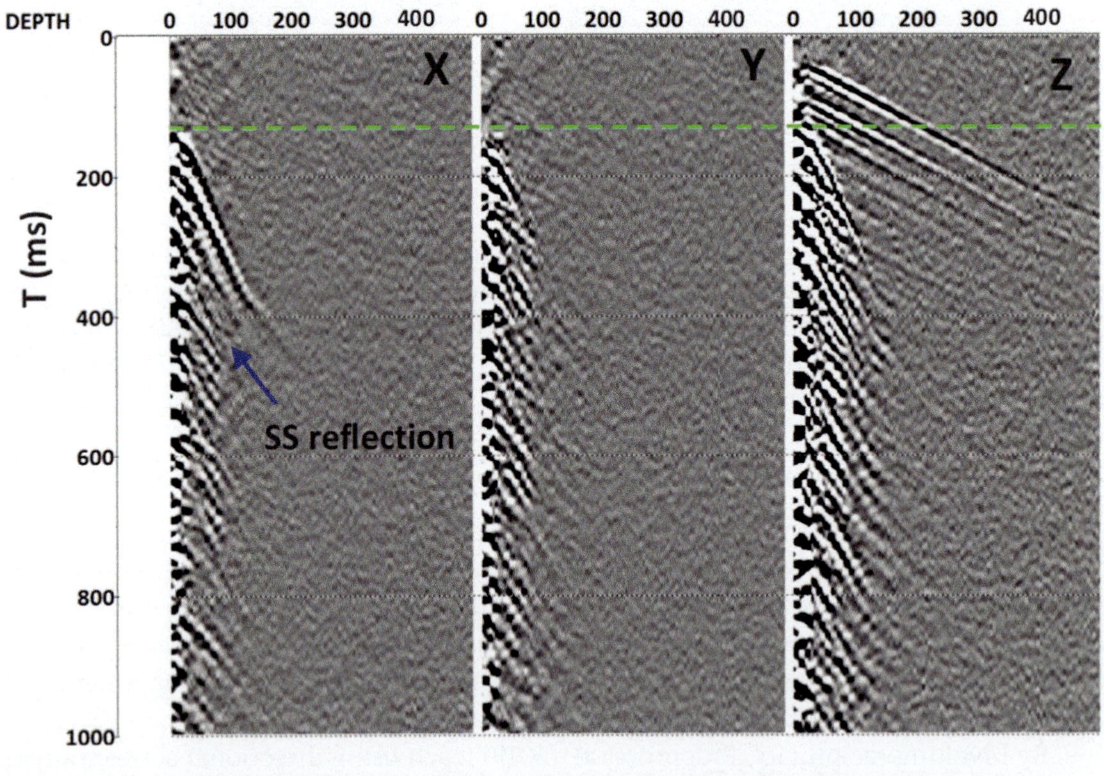

Figure 7.13. The 3C VSP DAS records acquired using a weight drop after Alford rotation into orthogonal directions. (Modified from Tertyshnikov and Pevzner, 2020, Figure 2.)

For applications such as VSP that use active sources, this method is beneficial. However, for passive applications such as earthquake monitoring, this obviously won't be appropriate because we have no control over repeating the source in different ways, multiple times. At an even deeper level, a three-component seismic survey is a subset of a nine-component seismic survey where all of the receivers are 3C and all of the sources are also oriented in all three directions. So, the solution in this section does not address

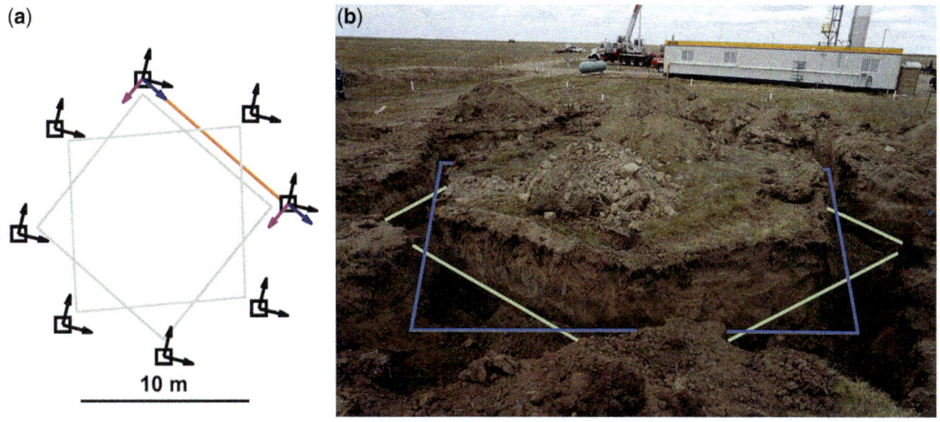

Figure 7.14. (a) Acquisition geometry for two fiber optic cable loops (gray) and eight 3C geophones (black boxes), and (b) photo of the trench showing each loop in a different color. (Modified from Hall et al., 2020, Figure 4 and Innanen, 2019, Figure 1.)

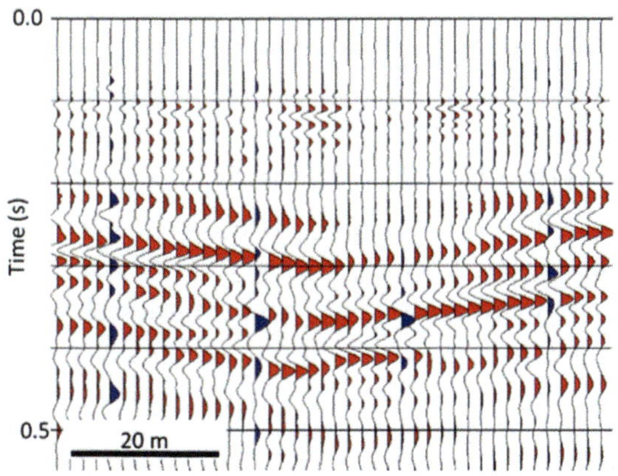

Figure 7.15. The DAS strain rate traces (red) for one of the squares shown in Figure 7.14, and the corresponding geophone traces converted to strain rate (blue). (Modified from Hall et al., 2020, Figure 6.)

the full 9C survey requirements, but it is a significant improvement over a one-component DAS survey.

7.6. Creating two-component horizontal surveys using buried directional DAS sensors

If the funding is available to bury fiber, not only in a trench but also in a more complicated arrangement than a straight line, then it is possible to acquire DAS data in different horizontal orientations. Innanen et al. (2019) and Hall et al. (2020) describe creating loops of fiber, each in the shape of a square, buried 2 m below the ground surface, as shown by the gray lines in Figure 7.14a. The small black boxes with arrows indicate geophones which were used to verify the DAS data. A photo of the trenched loops of fiber is shown in Figure 7.14b. Each side of the square must be larger than the gauge length in order to fully sample a continuous span in each direction. The notion is that the fiber on each side of the square loop will measure the strain wavefield oriented in the axial direction of that portion of the fiber. They have run two complete loops of fiber in each square so that the S/N could be improved by stacking together the traces from the corresponding coincident portions of the fiber.

Figure 7.15 shows the extracted DAS traces (in red) for one complete square loop and the geophone traces which have been converted to strain (in blue). For this example, the gauge length is 5 m and each side is 10 m, so there should be five traces in the middle of each side of the square which are fully oriented in the desired direction without any "wrap around" effect of the gauge "bending" around the corner of the square. The strain rate data from the geophones appear to match the DAS strain rate data quite well, thereby confirming the usefulness of this method.

7.7. Use of multiple fibers to extract the strain tensor

Instead of trying to obtain conventional 3C wavefield data, a completely different approach is to extract the unique 6 components of the full 9C strain tensor by using multiple optical fibers in a cable (Ning and Sava, 2016, 2018a, 2018b). Figure 7.16a, 7.16b, and 7.16c shows three different arrangements of fiber, respectively: two helical fibers with different wrapping angles, one straight fiber and one helical fiber, and a single fiber that changes its wrapping angle as a function of distance along the fiber.

Let's focus on the straight fiber with a second helical fiber wrapping around it, which I'll call cable type B (Figure 7.16b). Synthetic VSP data are created from the model in Figure 7.17a. The source is indicated by the red dot, and 500 m of cable type B is shown by the vertical black line. The synthetic, noise-free six components of strain are shown in Figure 7.17b. The objective is to invert the simulated DAS strain measurements from the straight and helical fibers and obtain a close match to the actual 6C strain values. The DAS data are simulated with a very small gauge length of 0.2 m, a channel spacing of 0.2 m, and a wrapping angle of 20°. The cable diameter is 0.0244 m (about 1 in).

We know that the gauge averages the strain in the fiber over the length of the gauge. Thus, a large gauge length will average out all of the angular variation in the strain, making it impossible to extract the azimuthal (and thus 6C) character of the wavefield. The proposed inversion method uses the notion that there is a relationship between the wrapping angle, the channel spacing, the cable diameter, and the gauge length that will allow the azimuthal variation in strain to persist in the measured DAS

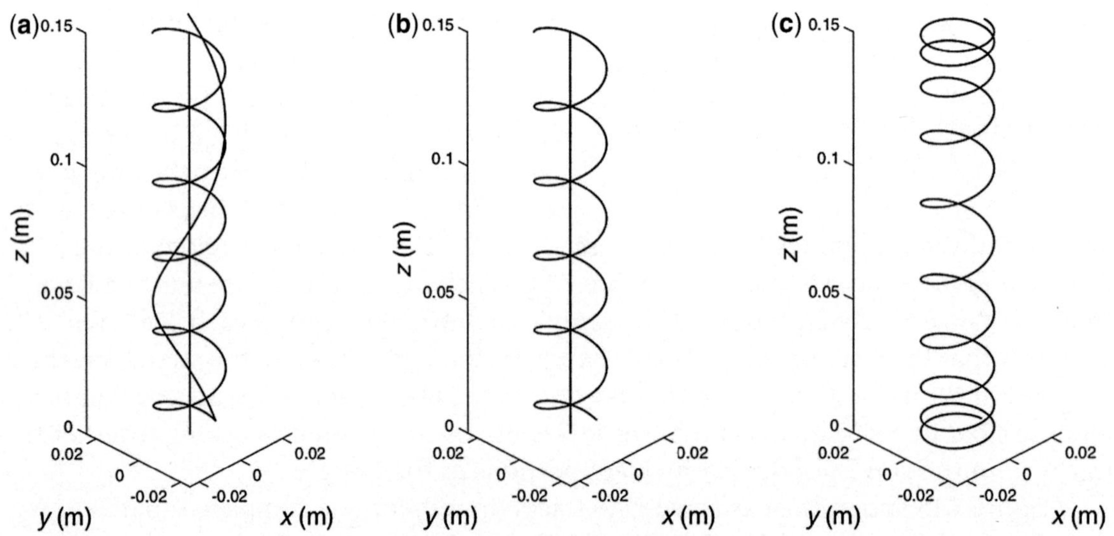

Figure 7.16. Wrapping geometries for three different fiber cables. (a) Two fibers with different wrapping angles. (b) A single straight fiber with a second fiber helically wrapped. (c) A single fiber that changes its wrapping angle with distance along the fiber. (Modified from Ning and Sava, 2018b, Figure 2.)

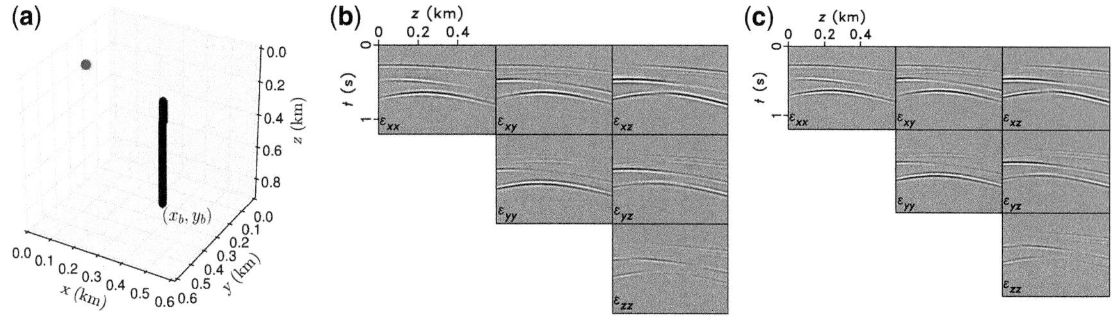

Figure 7.17. (a) Acquisition geometry for the synthetic VSP record. (b) Modeled 6C strain tensor. (c) Inverted 6C strain tensor using cable type B. (Modified from Ning and Sava, 2018b, Figures 4a and 5a.)

strain data, which then can be inverted to extract the 6C strain data. The result of their inversion is shown in Figure 7.17c, which matches the synthetic data very closely.

The upside of the approach is obvious – instead of struggling to obtain two or three components of strain using different acquisition geometries, 6C strain data can be acquired with a relatively simple design of a helically wound fiber optic cable. The challenge, however, is that very small gauge lengths are required, which introduces more noise into the raw DAS data. As the advancements in IU design provide for high signal-to-noise ratios for small gauge lengths, this method will, no doubt, be very important.

7.8. Summary

As we have seen, straight optical fiber allows us to measure a single component of strain. By employing helically wrapped fiber around a mandrill in the cable, we can expand the response of the fiber to be more like a hydrophone. However, care must be used when designing the cable so that an appropriate balance between P- and S-wave energy is preserved for the desired application.

Using an active directional source and energizing it in three directions can allow us to record 3C DAS data. For surface seismic applications, directional DAS sensors can be designed by burying loops of fiber in the ground, if there is free access to digging large form factor trenches. Finally, the use of multiple fibers designed appropriately in a cable may allow us to extract 6C strain measurements.

Chapter 8 Sources of noise in DAS data sets

The easiest way to avoid wrong notes is to never open your mouth and sing. What a mistake that would be.

— *Pete Seeger*

Perfection is the enemy of the good.

— *modified from Voltaire*

8. Sources of noise in DAS data sets

As geophysicists and engineers, we understand that there are many sources of noise in seismic data. Non-optical sources of noise include cultural and environmental noise. In DAS measurements, there are three types of unexpected "optical noise." In the VSP environment, flow noise can be bothersome when the well is in production.

8.1. Non-optical sources of noise

Cultural noise typically comes from sources such as: road noise from traveling vehicles, footsteps of people and animals, construction noise, airplanes, ships at sea, and electrical 60 Hz power lines. Environmental noise comes from sources such as: wind across the sensors, ocean waves crashing on beaches, earth tides, and earthquakes. Many approaches have been taken to mitigate these effects.

A simple in concept, but difficult in practice, way to reduce cultural noise is to install the fiber cable far away from cities, airports, construction sites, and roads. Another approach is to take measurements at night when cultural activities tend to be much less prevalent. Environmental noise is more difficult to avoid because it tends to be ubiquitous. For surface seismic and earthquake monitoring, burying the fiber optic cable may help reduce the wind noise.

As these approaches will likely not completely mitigate noise issues, there are many signal processing approaches to address them. If you are using an active source, then repeating the measurement multiple times and averaging the measurements is a classic way to increase S/N by stacking out random and non-repetitive noise. Using digital processing methods such as velocity filters may help remove seismic waves emanating from wind noise and vehicular traffic. Bandpass filtering also can remove unwanted types of noise and increase the S/N of the seismic signal.

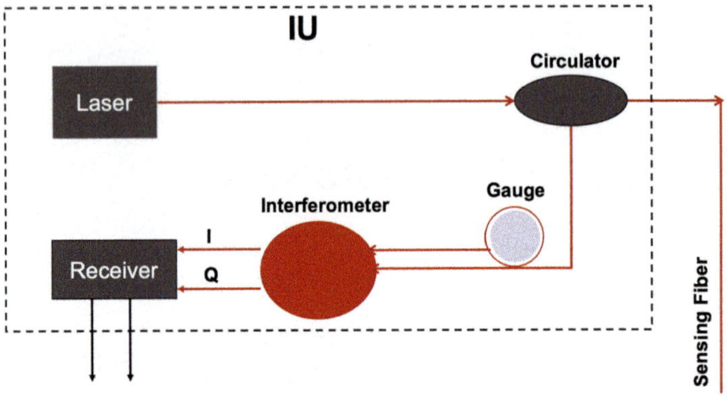

Figure 8.1. Schematic of an IU.

8.2. Common mode noise

The first type of optical noise is called common mode noise (CMN). As shown in Figure 8.1, the heart of the IU is the optical interferometer which takes the backscattered light from the direct path and the path through the gauge as inputs and brings them together to form the I and Q signals (defined and described in Section 3.9). Unfortunately, the interferometer itself is sensitive to sounds that enter the IU. Sonic waves that hit the interferometer slightly deform the two input arms and are imprinted as "infinite velocity" events on all of the strain (rate) data that is recorded (Jeff Bush, personal communication). So, one approach to reduce the amount of CMN on a record is to sonically isolate the IU to remove as much interfering noise as possible from being detected by the interferometer.

Figure 8.2 shows an example VSP record with CMN. The zone labeled "A" spans some of the surface cable and shallow well bore with multiple casings which greatly reduces the S/N of the seismic waves. The zone labeled "Vertical" is the vertical portion of the well which shows a large down-going P-wave arrival, a much smaller down-going S-wave, and various up-going P-wave reflections. The zone labeled "Lateral" covers the lateral horizontal part of the well where the seismic waves are much lower in amplitude. The zone labeled "B" contains channels past the end of the fiber, which is pure noise. Each trace is scaled by its maximum amplitude so that the smaller amplitude events at depth can be seen and not dominated by the large shallow events. The pervasive flat horizontal events that seem to start at about 500 ms are the CMN. This is sometimes

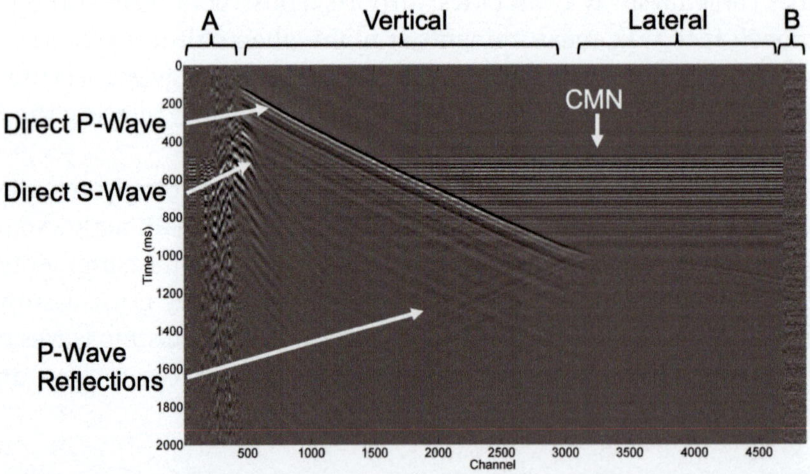

Figure 8.2. Example DAS VSP record with common mode noise (CMN).

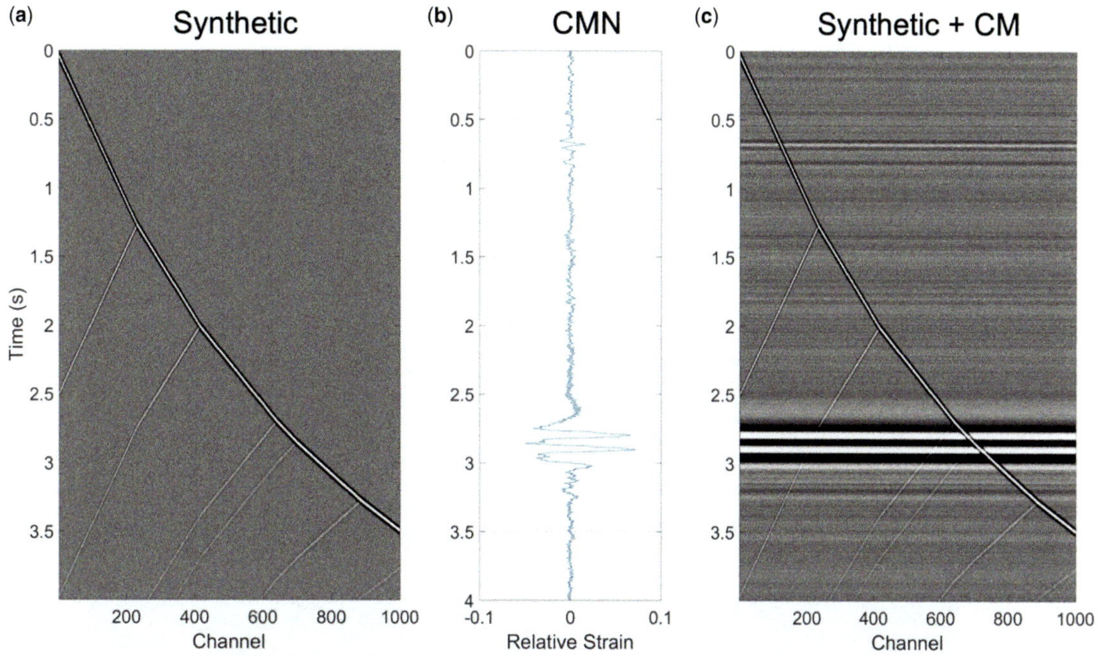

Figure 8.3. (a) Synthetic VSP record. (b) Actual CMN noise trace. (c) Synthetic VSP record with CMN added. (From Ellmauthaler et al., 2017, Figure 2.)

called horizontal noise because it appears as horizontal striping on a DAS record with distance along the fiber horizontally, and time vertically.

While CMN is annoying, it is not fatal, because it can be (mostly) processed out of the data (Willis et al., 2019a). To illustrate how this is done, let's create a synthetic example. Figure 8.3a is a six-layer synthetic VSP record to which I've added random noise. Figure 8.3b is a single trace recording of the CMN. I've added the CMN shown in Figure 8.3b to each trace in the synthetic record in Figure 8.3a. The resulting VSP record with CMN is shown in Figure 8.3c, where the bright bands of horizontal noise are clearly present. If all of the traces in a DAS record are stacked together, we expect that the signal will stack out because it is not consistent across the record. However, the CMN will stack constructively together to provide a good estimate of the CMN waveform. Thus, we can estimate the CMN noise by stacking together all of, or at least a significant portion of, the DAS record, and subtracting that estimate from each trace in the DAS record.

Let's now look at a field data example. Figure 8.4a shows the VSP record from Figure 8.2 without all the annotation. I have summed all of the traces in this VSP record together to form an estimate of the CMN. Then, I've subtracted the CMN trace from each of the traces in the VSP record. The result is shown in Figure 8.4b. Most of the CMN has been removed and, because each trace is scaled by its maximum, we can now see the much lower amplitude seismic events in the lateral portion of the well (channels 3250 to 4700).

It should be noted for data acquired with an active source that is repeated multiple times, each individual recording can have this process applied separately. Alternatively, if a simple stacking process is used, it will be a linear system, so the CMN removal process can be applied to the data after stacking.

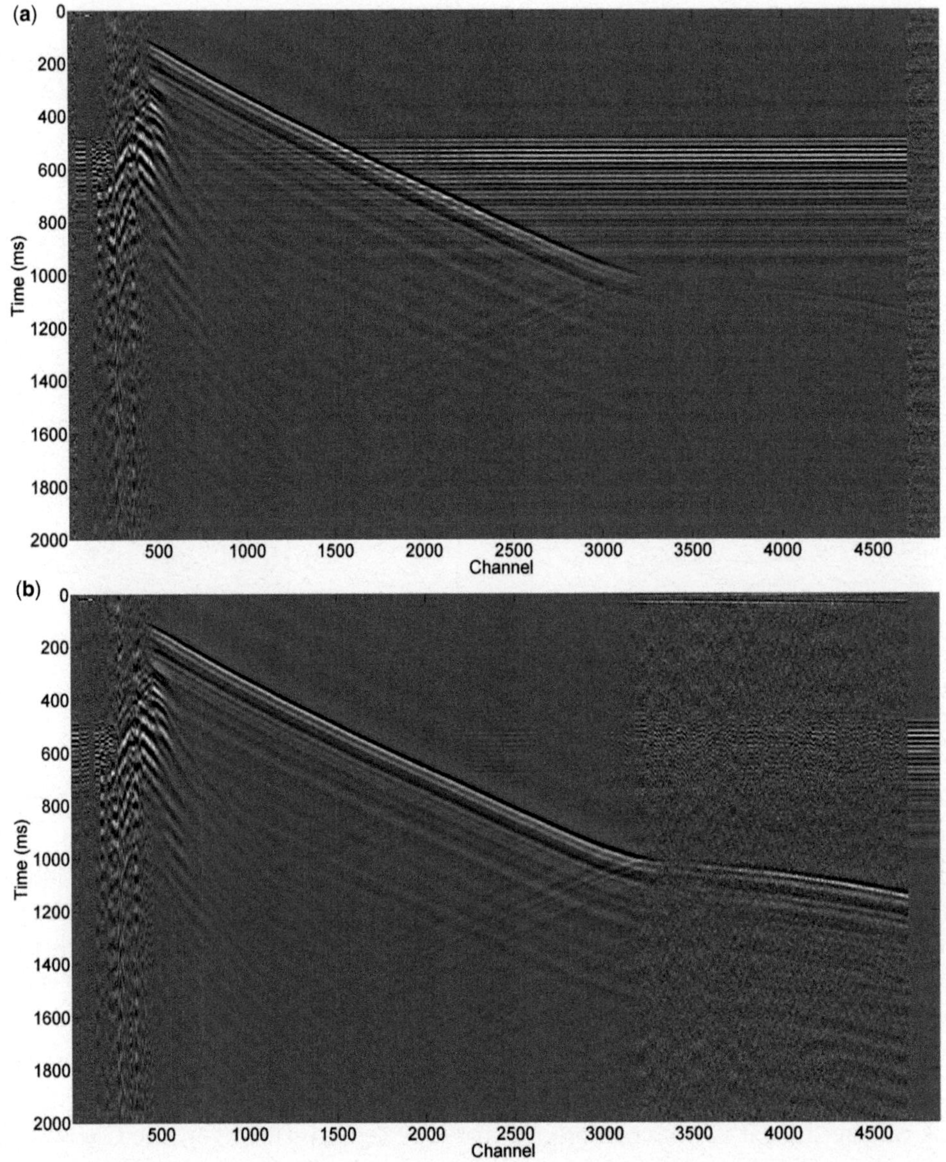

Figure 8.4. (a) The VSP DAS strain rate record showing CMN and (b) the same record after processing to reduce fading.

8.3. Optical fading

The second source of optical noise is called fading. This is a seemingly confusing name because the word "fading" seems to imply a small effect, but the actual expression of fading is extremely large amplitude noise bursts on the seismic (strain) data. It also is called vertical noise because, on a DAS record with time on the vertical axis and distance along the horizonal axis, the noise appears as vertical streaks on portions of individual traces. Fading occurs in all DAS systems because the laser light is interacting with an intrinsic property of the optical glass. In fact, it has been estimated that 3% of the fiber is faded at any one moment in time (Ellmauthaler et al., 2016b). Let's first talk about the physics of what is causing this noise, show some examples, and then discuss how to mitigate it.

The basis of a DAS measurement comes from light that is backscattered from random small changes in the index of refraction at sites all along the sensing fiber. In addition to sound vibrations, changes in the temperature of the fiber make minute changes in the shape of the fiber over all time scales, which slightly shift the scattering sites in the fiber. The amplitude of the light that is backscattered from these sites may be large or small. The DAS measurements essentially ignore the amplitude of the light and only use the phase of the difference of the light between places separated by the gauge length. Provided that the amplitude of light is large enough, we can use equation 3.19 to estimate the phase angle, θ, between the I and Q signals (described in Section 3.9 and repeated here) as

$$\theta = \operatorname{atan}\left(\frac{Q}{I}\right) \tag{8.1}$$

The length of the vector that is formed by I and Q is called the radius, given by

$$radius = \sqrt{I^2 + Q^2} \tag{8.2}$$

is an indication of the amplitude of the backscattered light. The aim is to have the radius equal to unity. If the radii of the I and Q signals are large compared to any noise

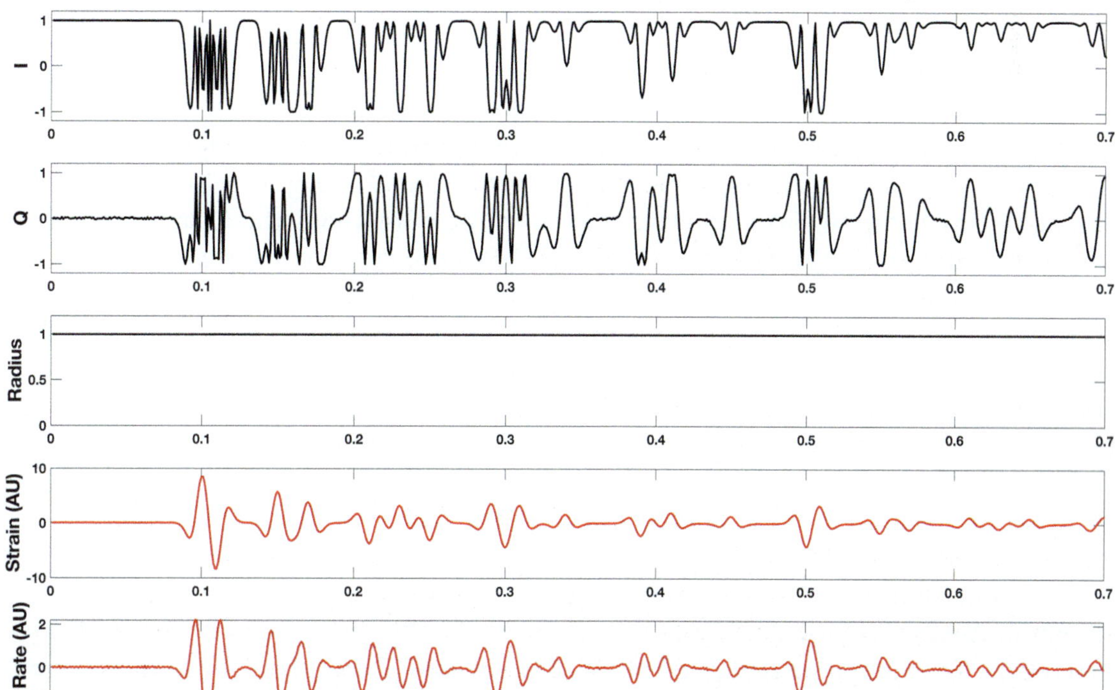

Figure 8.5. Synthetic example to demonstrate extracting strain and strain rate from the I and Q values.

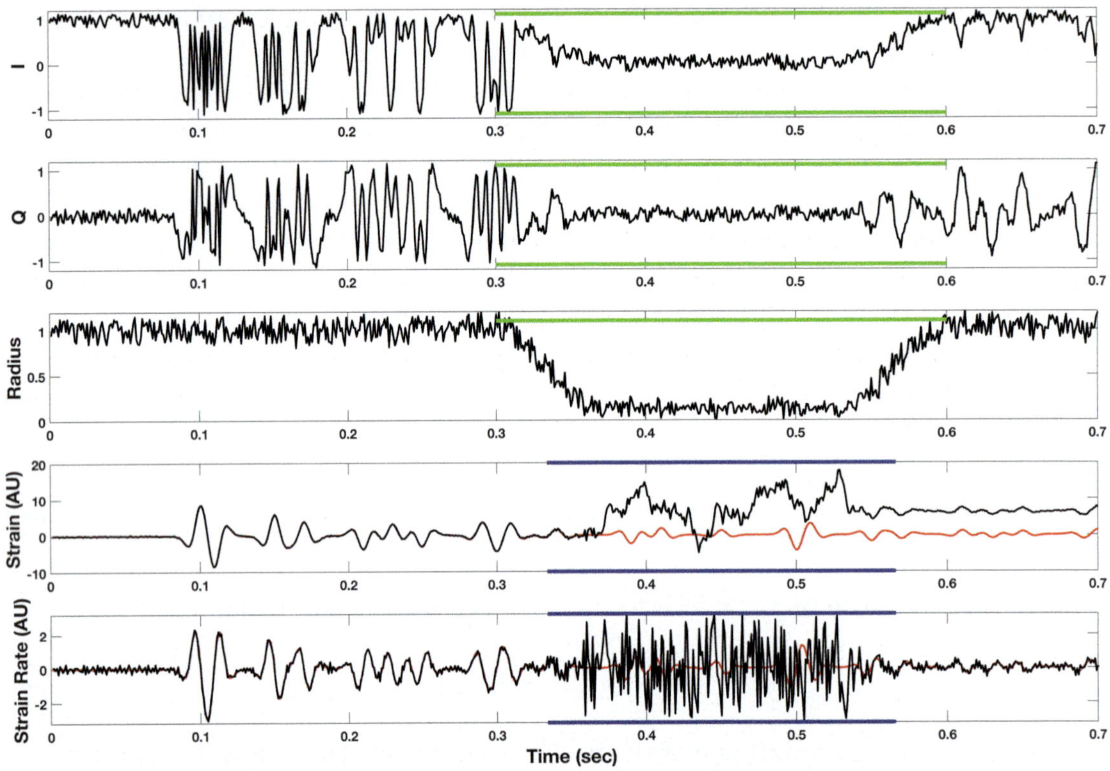

Figure 8.6. Synthetic example demonstrating fading between 0.325 and 0.575 s.

in the signals, it is easy to extract a reliable estimate of the phase angle. However, whenever the amplitude of the backscattered light is small, noise will dominate the arctangent operator and create large erroneous swings in the phase computation.

Let's look at a synthetic example which is fashioned after Figure 1 in Ellmauthaler et al. (2017). The top two rows in Figure 8.5 show the noise-free I and Q signals for a 0.7 s time window of DAS data. The third row shows the radius computed from the I and Q values, which is a constant unity amplitude in this example. The fourth row shows the unwrapped arctangent of I and Q. It is labeled as strain, but it is actually the phase. Using equation 3.18, this could be converted to strain by multiplying it by a constant, but for our purposes here I will label it as strain with arbitrary units (AU). The bottom row shows the corresponding strain rate which is simply the time derivative of the strain.

Now let's add some noise to the I and Q values, and then introduce fading. Random noise is added to the I and Q traces individually. Then, they are multiplied by a tapering function (1-0.85*tukeywin from MATLAB). The resulting I and Q values are shown in the top two rows of Figure 8.6, where I've highlighted the faded zone with the green bars. The computed radius in the third row clearly shows than the radius drops to nearly zero between 0.36 and 0.54 s. A polar plot of the I and Q values for a non-faded zone, between 0.15 and 0.25 s, are shown as black dots in Figure 8.7a. The dots hug the (blue) unit circle indicating that good values of phase can be computed from them. The black dots in Figure 8.7b show that the I and Q values from the faded zone, between

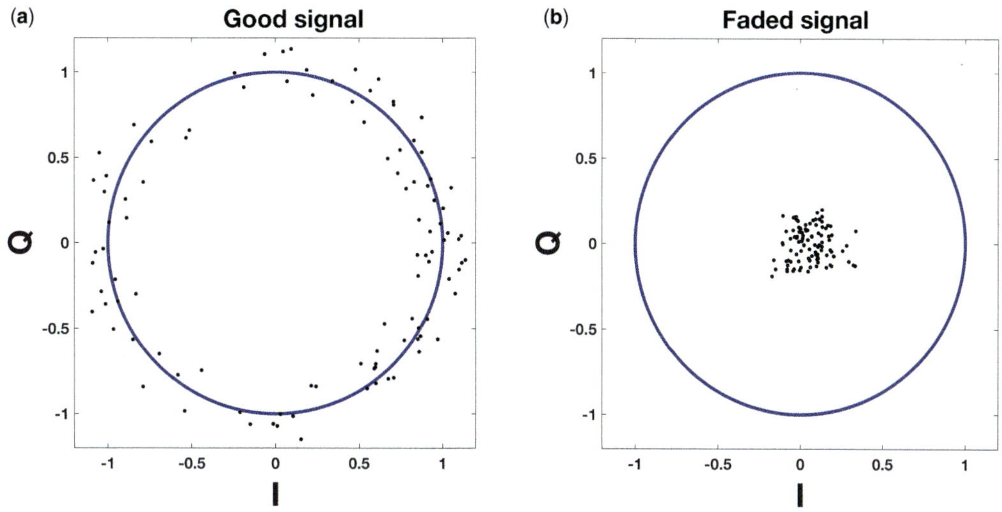

Figure 8.7. The I-Q circle showing (a) good signal strength between 0.15 and 0.25 s and (b) faded signal between 0.35 and 0.45 s.

0.35 and 0.45 s, fall close to the origin. This means that small values of noise can easily move the phase into the wrong quadrant and introduce large errors.

As we did for the noise-free case, the computed strain from the noisy faded values is shown by the black line in the fourth row of Figure 8.6. The noise-free case is shown by the red line. The blue bars show the zone where the effects of fading seem to be significant. We see two major effects. First, there are seemingly large changes in the waveform in the faded zone. These are caused by the inability of the arctangent operator to pick the correct phase angle which introduces large jumps. Second, we see that, after the fading has stopped, there is a DC shift in the trace. For times larger than 0.55 s, the faded strain trace (black) is about 7 units higher than the noise-free case (red). This is caused by the inability of the phase unwrapping algorithm to determine the appropriate multiple of π to add to the trace values in the faded zone.

The bottom row in Figure 8.6 shows the strain rate values for faded data in black and the noise-free data in red. The derivative operator has removed the DC shift between the faded and noise-free data at the end of the trace. It also has changed the jumps in the strain trace to being large spikes in the strain rate trace. Thus, fading affects both strain and strain rate data with detrimental outcomes.

The approach to mitigate this effect has evolved over the years with three different solutions. The simplest approach is to perform despiking operations on the strain rate data using signal processing (Bakku, 2015). The notion is to detect large spikes and either mute them out, interpolate through them, or taper the data in that region. While practical, it is not very satisfying because this leaves gaps or poor data in that zone.

The second approach is to repeat the measurement one or more times and then attempt to choose the good data over the faded data when stacking it together. This works because, over time, the location of the fading changes both along the fiber and in time on the trace.

The third approach is to enhance the IU to have more than one laser, each tuned to a different wavelength of light (Hartog et al., 2016). The spatial and temporal location

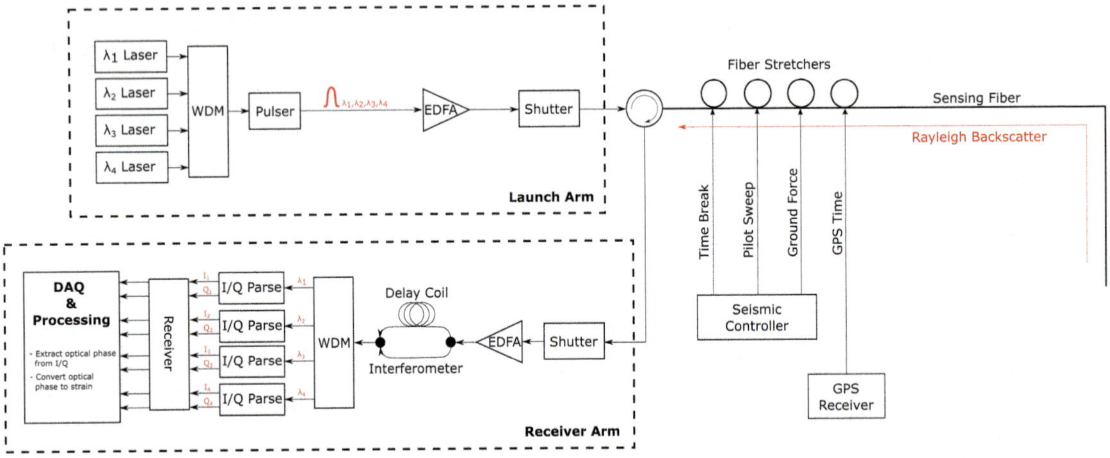

Figure 8.8. Schematic of a quad-laser IU. (From Ellmauthaler et al., 2020b, Figure 5.)

in the DAS record of fading for each laser will be different based upon how the wavelength of each light interacts with the scattering points in the fiber. Figure 8.8 shows a complicated schematic of an IU with four lasers which allows for four simultaneous estimates of the strain (rate) to be created. Then, these four estimates can be combined using weighted stacking to exclude the faded portions of the data and save the desired signal (Daley et al., 2016; Ellmauthaler et al., 2017). In many ways, this is just another way to repeat the measurement, but without consuming more acquisition time.

Figure 8.9 shows four DAS VSP strain rate records from two acquisition sweeps using an IU with two lasers. Faded traces are shown by the high amplitude traces that disrupt the coherent VSP signals. Notice that the locations with fading are in different trace locations between the four records, so the challenge is how to combine these four

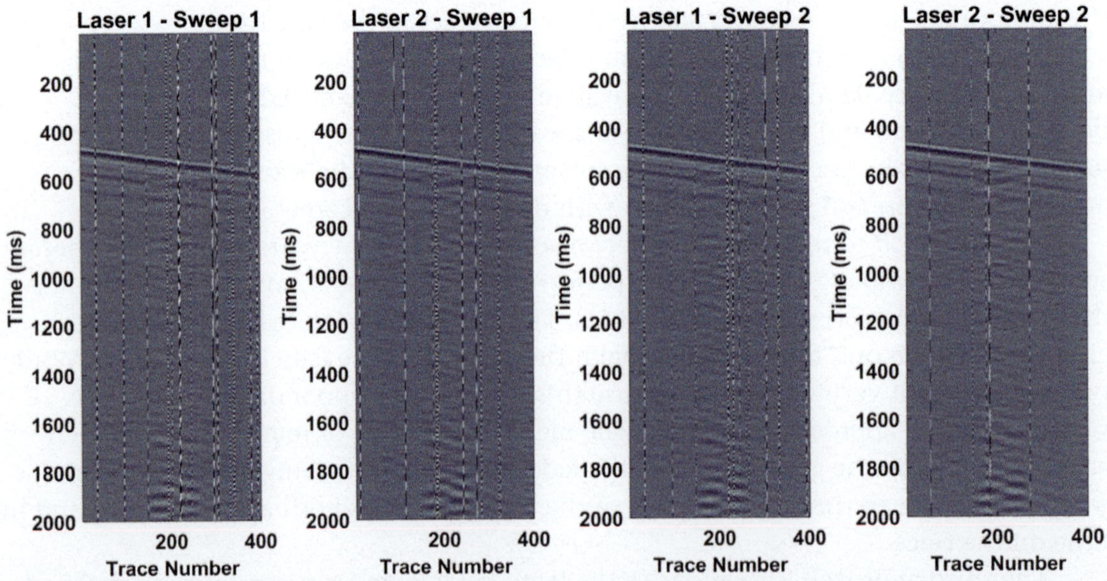

Figure 8.9. Example of four DAS strain rate records from two acquisition sweeps and two lasers.

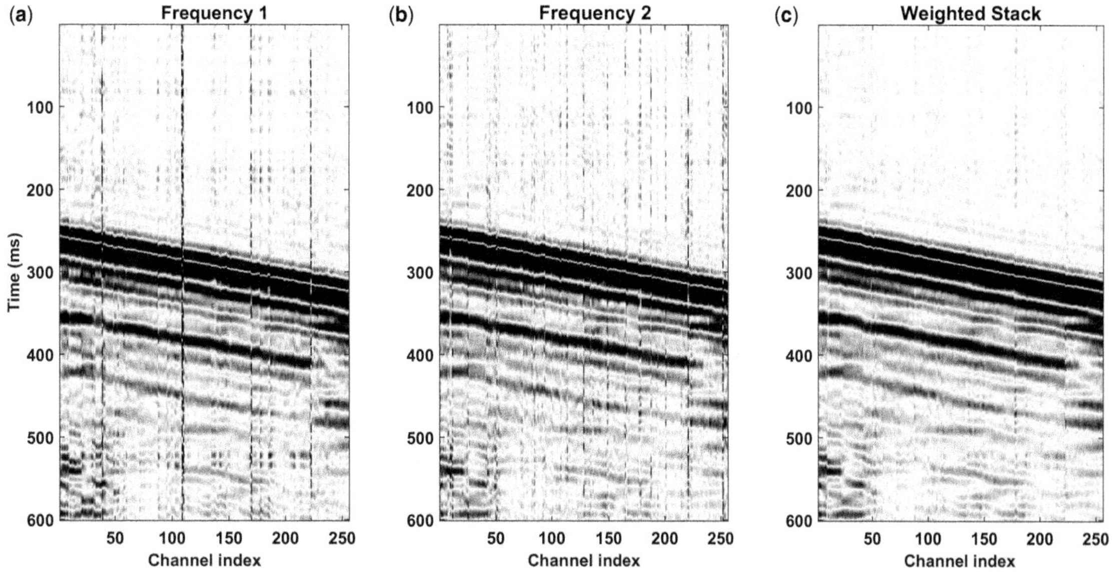

Figure 8.10. Example of weighted stacking. (a) Weighted stack of two sweeps for laser 1. (b) Weighted stack of two sweeps for laser 2. (c) Weighted stack of records from both lasers. (From Willis et al., 2022, Figure 7.)

records in a way that excludes the high amplitude faded traces and enhances the good signal. A simple approach is to use median stacking assuming that there are enough good traces for each channel. Unfortunately, median filtering is computationally expensive. Weighted stacking is much quicker and yields very good results.

Figure 8.10 shows a magnified view of the weighted stack of these data. The first column is the weighted stack of both sweeps for laser 1. The second column is the weighted stack of both sweeps for laser 2. The third column is the weighted stack of all the data. It is clear that most of the spiky faded traces have been mitigated and replaced with good signal data.

8.4. Temperature sensitivity

The DAS strain (rate) response is sensitive to both sound vibrations and temperature. Bakku (2015, Section 3.4) states that the temperature response of the fiber optic glass comes mostly from an induced change in the index of refraction. It typically affects the spectrum of the DAS data below 2 Hz, so he suggests that it is possible to remove the temperature response by a low-cut filter applied to the DAS strain record. However, for applications such as low-frequency strain measurements of hydraulic fracturing, a simple low-cut filter will likely remove the effects that are to be monitored.

Ellmauthaler et al. (2017) show an example of DAS VSP data acquired with a permanent fiber optic cable cemented behind casing. Contemporaneous with the acquisition of the DAS VSP data, geophone VSP data also are collected. The top panel in Figure 8.11 shows uncorrelated relative strain data from eight of the vibrator sweeps. Each of the traces represent the DAS data acquired during a unique clamp location of the geophone tool string. All eight traces show a rising to the right, low-frequency trend while the seismic signal is very high frequency by comparison riding on the trend. Two

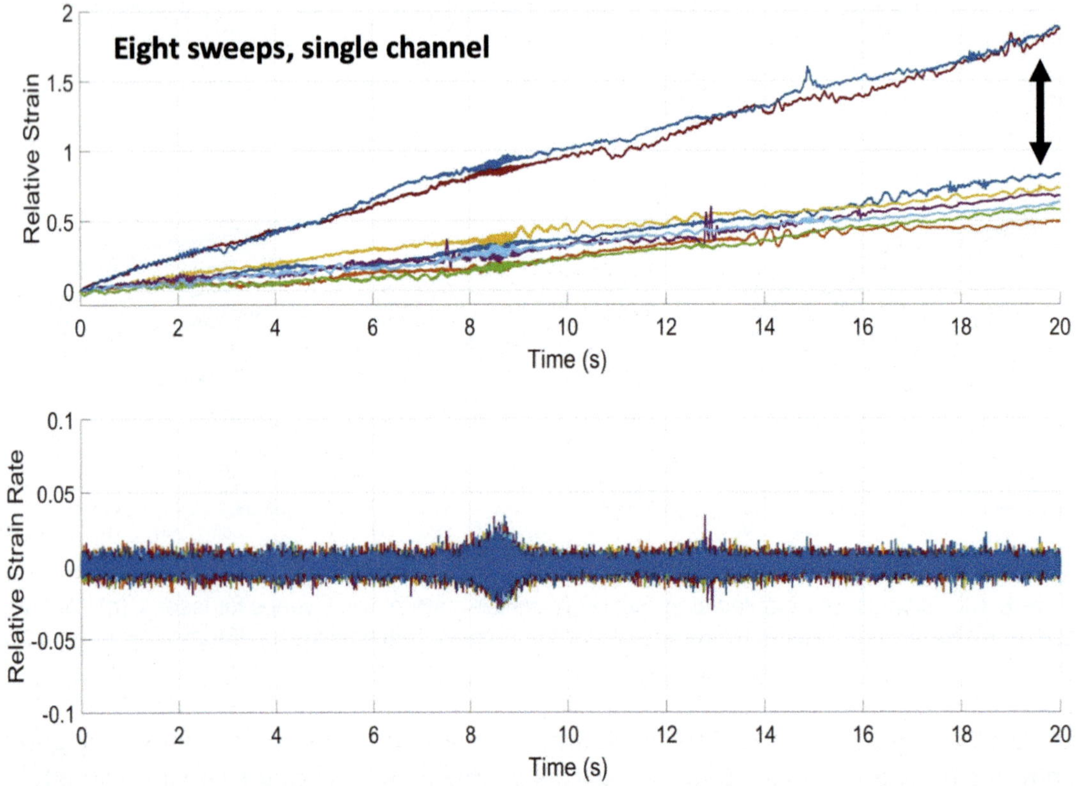

Figure 8.11. The VSP traces from a single channel and eight sweeps: relative strain (top) and strain rate (bottom). (Modified from Ellmauthaler et al., 2017, Figure 3.)

of the traces have trends which are much steeper in slope. These two traces have been acquired while the geophone tool string was in the vicinity of that channel.

The low-frequency trend in all of these traces appears to be much lower than 2 Hz, and over a 20 s time period appears to be relatively linear, implying a steady change in temperature. Lowering the geophone tool string into the borehole has disturbed the temperature profile in the borehole fluid. In addition, it appears that, when the geophones are positioned near the channel of the fiber we are observing, they are imparting additional temperature change.

The bottom panel in Figure 8.11 shows the result of taking the time derivative of the eight relative strain traces to convert them to strain rate. A simple derivative has removed the low-frequency drift from the data. This is not surprising because a derivative operator is equivalent to multiplying by $-j\omega$, which greatly attenuates low frequencies and enhances high frequencies.

8.5. Poor coupling in VSPs

A unique source of noise is encountered for DAS VSPs in vertical wells using either wireline or disposable fiber. While the exact physics are not well understood yet, it is believed that in vertical wells the fiber optic cable is not in contact with the borehole wall over portions of the well. In areas with no coupling, the cable appears to resonate back and forth when the seismic energy hits the cable, thereby forming zigzag noise

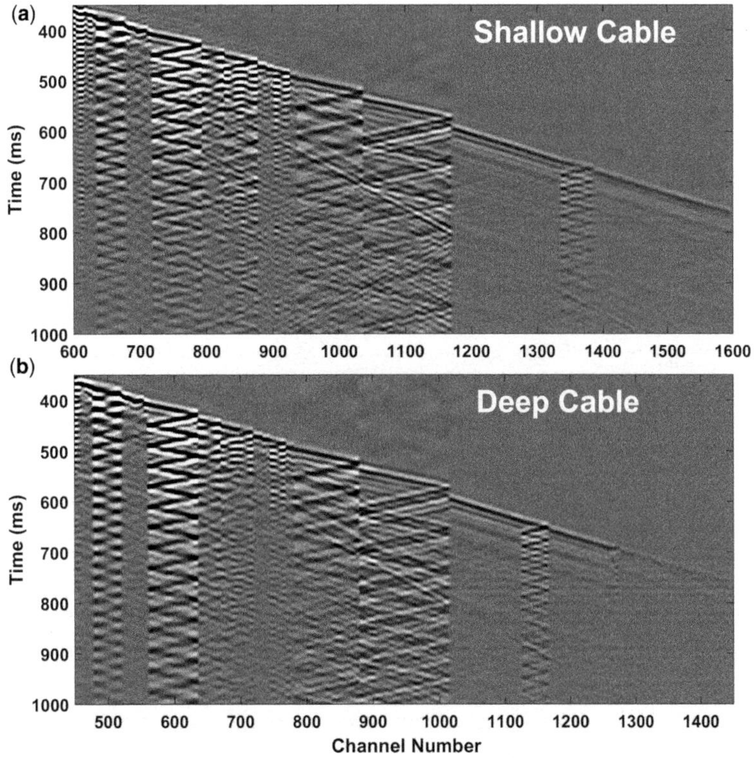

Figure 8.12. Examples of zigzag noise in the same well. (a) A DAS VSP record with the wireline cable pulled up to a shallow location, and (b) the DAS VSP record with the cable deeper. (From Willis et al., 2019b, Figure 1.)

patterns (Cai et al., 2016; Dean et al., 2016b; Willis et al., 2019b). The occurrence of this noise is most prevalent for small-offset source locations. Far-offset sources do not seem to excite this noise.

Figure 8.12 shows DAS VSP data acquired in a vertical test well using a wireline fiber optic cable. Figure 8.12a is a record when the cable was not lowered all the way to the bottom of the well, while Figure 8.12b is when the same cable was lowered an additional 150 m deeper. Many zones of the data show zigzag noise patterns that we have come to associate with the poor cable coupling issue. The two records are aligned so that the same depth interval is covered, making the channel numbers offset by 150. It is clear that much of the records are very similar – many of the zigzag patterns have repeated. However, there are zones that are quite different. Observe the zone around channel 1375 in the upper record has zigzag noise, but the corresponding zone in the lower record (around channel 1225) is free from the noise. Look at the zone around channel 1150 in the lower record which has lots of zigzag noise, while the corresponding zone (around channel 1300) is totally free of noise.

These observations suggest that the source of the zigzag noise is not a defect in the cable itself, because by moving the cable most of the noise repeats at the same depth in the well. Because some of the zigzag patterns shift or disappear when the cable is moved, this suggests that it is not a feature of the wellbore itself. So, we can conclude that the relationship between the cable and wellbore is key. By moving the cable, some

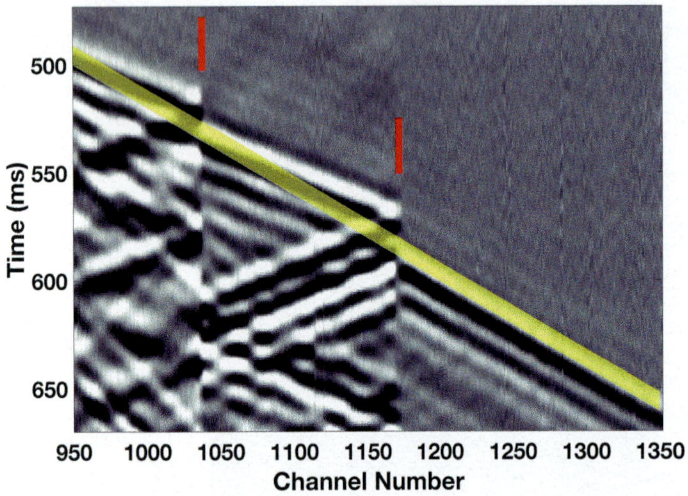

Figure 8.13. Magnification of Figure 8.12a with yellow line indicating the trend of the P-wave first arrivals. The red lines indicate breaks caused by the zigzag noise. (Modified from Willis et al., 2020b, Figure 2.)

of those relationships stay the same and the zigzag noise repeats, and some of those relationships change which in turn changes the zigzag patterns.

Why is the zigzag noise a problem? There are two major reasons. First, it is common for the first arrivals time picks on the record to be used to determine the interval velocities on a zero-offset VSP. When the source is located near the well head, the direct P-wave arrival comes to the receivers and can be used to invert for the interval velocity of the formations all along the well. The first arrival time picks from offset shot records also can be used with tomography to derive the velocity away from the well. Figure 8.13 is a magnified view of Figure 8.12a. The semi-transparent yellow line depicts the trend of the P-wave first arrivals from the interval between channels 1174 and 1350. The trend does not match the arrivals in the upper channels which have the zigzag noise. If the abrupt breaks in the first arrivals, shown at the red lines, were used in the velocity inversion algorithm, the result would be non-physical, extremely slow velocity breaks in the profile which are not related to the formation properties.

The second reason is that the up-going reflection events used for creating migrated images of the subsurface have been contaminated by artificial non-geologic up-going events. These events will create artificial breaks in the migrated images which are non-physical.

What can be done to mitigate the zigzag noise issue? One simplistic approach is to not log vertical wells with wireline and disposable fiber. Of course, many times that is not a practical solution. One proposed mitigation method is to avoid acquiring data with poorly coupled cables by adjusting the tension (slack) in the wireline cable (Frignet and Hartog, 2014; Constantinou et al., 2016; Schilke et al., 2017). Figure 8.14 shows the effect of slacking the cable on the coupling of the wireline to the wellbore. Under full tension (no slack), no usable data is acquired in this example. Progressively adding more slack to the cable increases the coupling to the wellbore and reduces the zigzag noise, until eventually the entire interval has good signal quality. Slacking the cable creates a helical winding of the wireline, pushing it against the borehole and creating coupling. A potential caution might be to make sure that the extra wireline cable in the borehole doesn't get stuck in the hole.

The other approach is to attempt to identify and remove the zigzag noise post-acquisition through signal processing techniques (Cai et al., 2016; Chen et al., 2018,

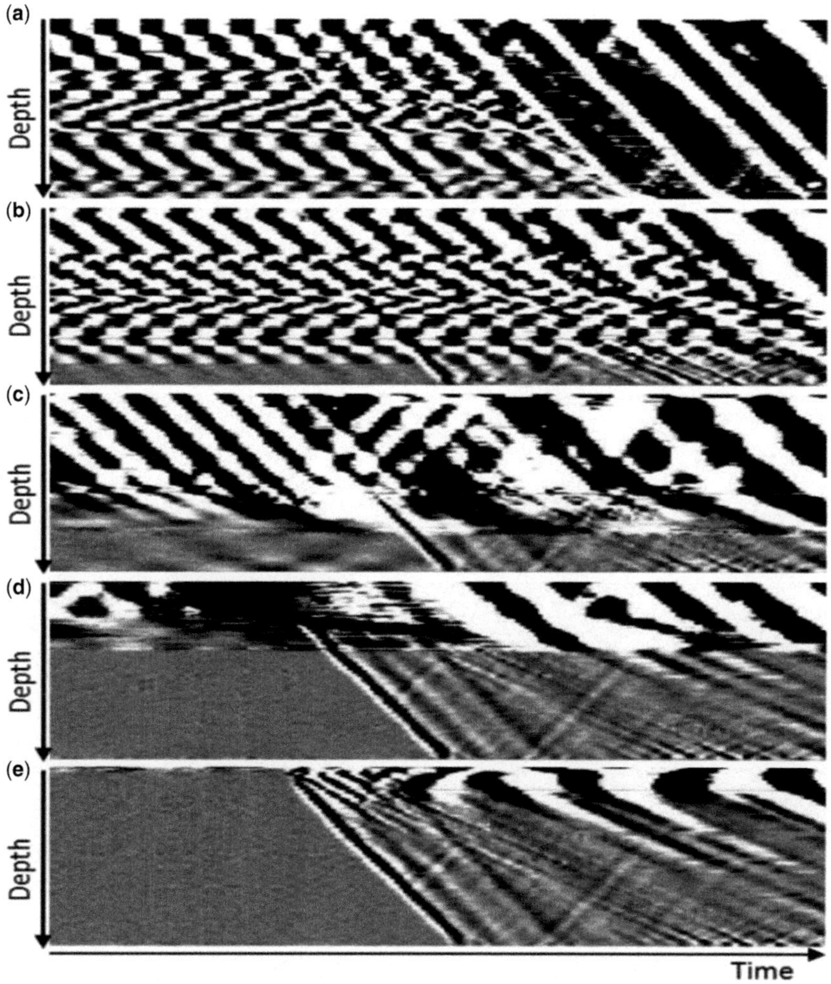

Figure 8.14. The DAS seismic data acquired with differing amounts of cable slack. (a) Cable under full tension. (b) Cable has 0.1% slack. (c) Cable has 0.5% slack. (d) Cable has 1.0% slack. (e) Cable has 2.0% slack. (From Constantinou et al., 2016, Figure 3.)

2019, 2020, 2021; Yu et al., 2018; Willis et al., 2019b, 2020b; Martuganova et al., 2019, 2021; Hou et al., 2021; Padhi and Willis, 2021). The properties of the zigzag noise are not well understood yet. The apparent velocity of the down-going (zig) and up-going (zag) noise patterns can be different from each other. In fact, they can change with depth in the well, ranging from 2500 to 4000 m/s (Willis et al., 2019b). This makes it challenging to model because it is unknown at this point how to predict their propagation characteristics.

Let's take a set of very simple synthetic examples to help understand the effect of zigzag noise on a data set. I'm fashioning these after work presented in Willis et al. (2019b). Let's start with a single zig down-going wavelet. Figure 8.15a shows 100 traces with a simple 60 Hz Ricker wavelet moving out across the record at 3600 m/s. Their corresponding spectra are shown below it in Figure 8.15e. In Figure 8.15b, I've added an up-going zag wavelet which is propagating at 3300 m/s. (The exact velocities are not important here because we are only trying to understand the effect of a pattern of

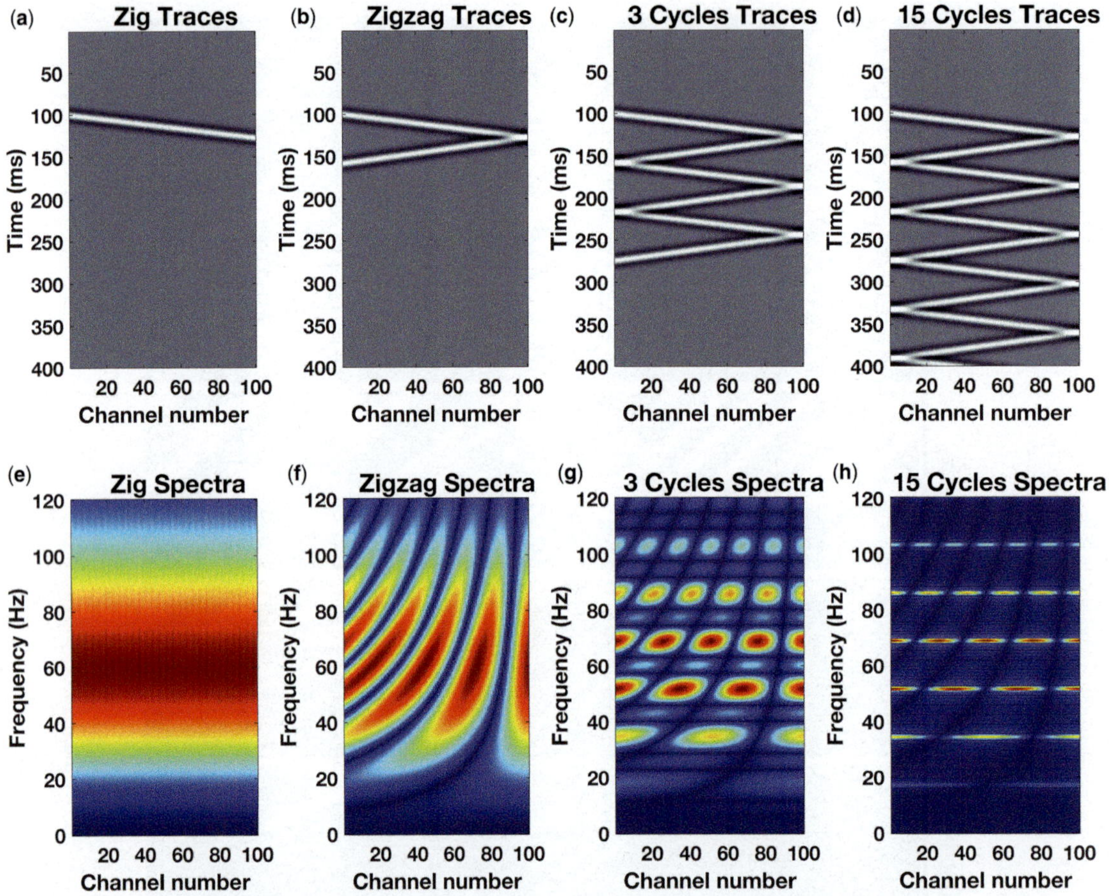

Figure 8.15. (a–d) Examples of traces with zigzag noise and (e–h) their corresponding spectra.

zigzag noise.) Figure 8.15f shows their corresponding spectra. These spectra show a significant change from the spectra of the simple 60 Hz wavelet for the single event, which might be surprising to you. However, from a basic understanding of signal processing, we know that two wavelets separated in time on a trace cause their spectrum to show peaks and notches related to that time separation.

In Figure 8.15c and 8.15d, there are 3 and 15 complete zigzag patterns, respectively. Their spectra are shown below them. The location of their notches and peaks are given by equations 1 and 2 from Willis et al. (2019b):

$$f_{peak} = \frac{p}{\Delta t} \tag{8.3}$$

$$f_{notch} = \frac{(n-1)+0.5}{\Delta t} \tag{8.4}$$

where f is the frequency of either the peak or notch, Δt is the time delay between the up- and down-going events, p is a positive integer starting at one representing each peak, and n is the integer order of the notch starting with one. There are additional resonances

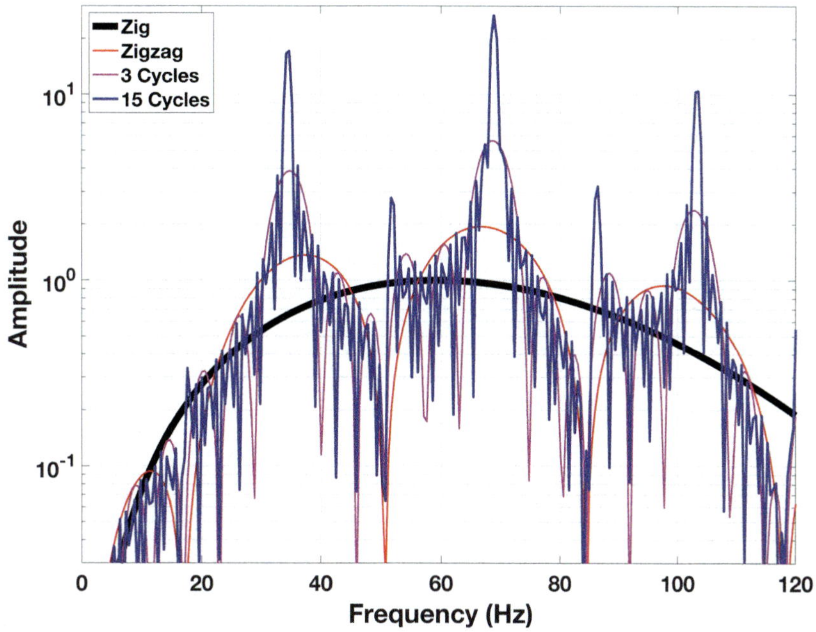

Figure 8.16. Amplitude spectra of channel 50 for each of the four cases shown in Figure 8.15.

from the repeated zigzag patterns, but equations 8.3 and 8.4 give the first order effects. Martuganova et al. (2021) show additional spectral analyses for zigzag noise.

We can see from this analysis that the zigzag noise imparts a predictable set of resonances in the spectra. It has been suggested that the zigzag noise is high frequency, but in fact each wavelet is quite similar to the incident P-wave. It is simply repeating that wavelet multiple times, which when separated by a time delay adds additional energy to the spectrum and creates notches and peaks. Figure 8.16 shows the amplitude spectra from channel 50 for each of the cases in Figure 8.15. The black line shows the spectrum for the single wavelet or "zig" case. The single zigzag cycle is shown by the red line, three cycles by the magenta line, and 15 cycles by the blue line. As we increase the number of cycles of zigzag noise, the spectra increases in energy because there is more signal in the trace. However, it is also obvious that these effects cannot be removed by a simple high-cut frequency filter.

The literature suggests several different signal processing methods to remove zigzag noise, including using a wavelet and DCT dictionary (Chen et al., 2018), sparse optimization (Chen et al., 2019), morphological component analysis (Chen et al., 2020; Hou et al., 2021), time-frequency domain filtering (Martuganova et al., 2019), adaptive Berg deconvolution (Martuganova et al., 2019, 2021), modeling and subtracting the noise (Yu et al., 2018; Martuganova et al., 2021), reverse deconvolution and adaptive subtraction (Willis et al., 2020b), and wavelet transform (Padhi and Willis, 2021). Many of these methods appear to be helpful and it will be interesting to see which of them become the workhorse algorithm(s). Figure 8.17 shows an example of zigzag denoising using the modeling and subtraction process by Martuganova et al. (2021). Figure 8.17a shows the VSP record with zigzag noise. Figure 8.17b shows the modeled noise. Figure 8.17c shows the record after denoising, which has clearly removed most of the noise.

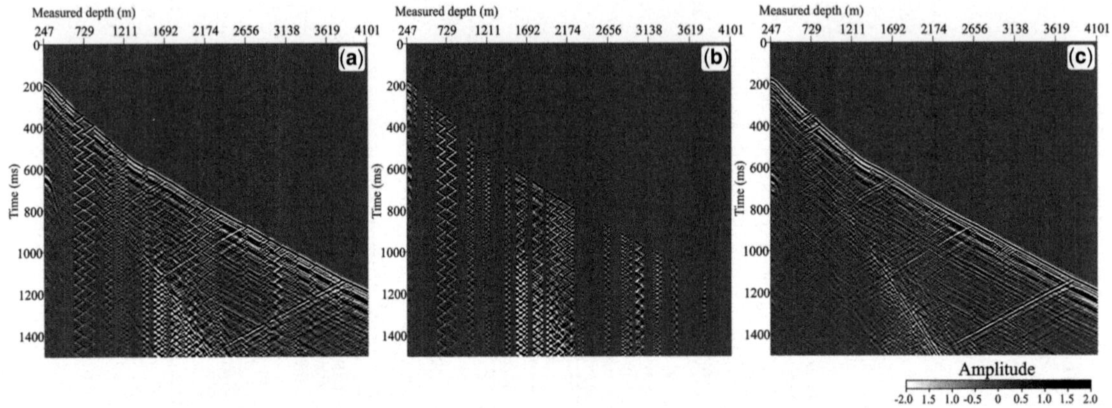

Figure 8.17. (a) A VSP record with zigzag noise. (b) Modeled zigzag noise. (c) Denoised record. (From Martuganova et al., 2021, Figure 9.)

8.6. Flow noise

Of great concern for acquiring VSP data to monitor time lapse changes in the reservoir is the noise created by production flowing in the well. Fluid flowing in the well bore with a fiber optic cable causes noise in DAS strain measurements. The amplitude of the noise varies with time and depth in the well.

Jiang et al. (2016) show flow noise in the range of 5–30 Hz on DAS VSP data acquired in the Valhall Field, offshore Norway. Saxon et al. (2018) show that production noise on a DAS VSP survey in an offshore Clair field well, west of Shetland, has broad spectrum noise that was particularly large, between 10 and 18 dB higher than data acquired without production flowing, in the 10 to 40 Hz band. Figure 8.18 shows an example of flow noise in a well in the Culzean Field, in the UK Central North Sea.

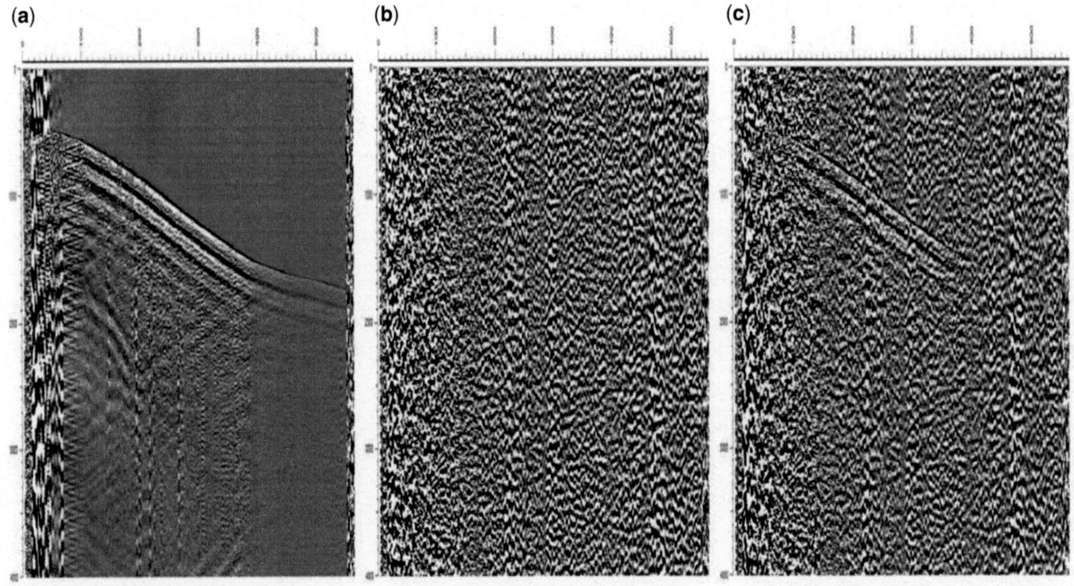

Figure 8.18. North Sea VSP record. (a) acquired without flow in the well, (b) single record taken while flowing, and (c) stack of two records taken while flowing. (From Merry et al., 2021, Figure 4.)

Ali et al. (2021) used tau-p and *f-x* infinite impulse response dip filtering to remove production noise, which was prevalent up to 35 Hz and spread out over a range of velocities in the shot domain. In the marine environment, it is impractical to repeat source points to stack out the random flow noise. Instead, high-fold data needs to be acquired and filtered in different data sorting domains so that the noise can be mitigated.

8.7. Summary

When acquiring DAS data, there are sources of noise that are common to all seismic data sets and then there are optical sources of noise that are unique to DAS data sets. Common seismic sources of noise are cultural and environmental, and decades of effort expended to address those sources for conventional geophone data sets now can be easily applied to DAS data.

One source of optical noise, called common mode noise (CMN), comes from sound around the IU imprinting on all of the DAS strain (rate) channels. This is relatively straightforward to remove. A second source of optical noise is called fading. This noise originates from the temporal and spatial changes of the fiber interacting with the wavelength of laser light. The result is that about 3% of all of the DAS data is contaminated with high amplitude noise bursts. This noise can be mitigated by either repeating the DAS acquisition measurements several times or adding additional lasers to the IU. Both of these solutions provide multiple estimates of the same data so they can be combined with a smart stacking operation.

A third source of optical noise comes from the sensitivity of the fiber to temperature changes. Extremely small changes in the temperature of the fiber all along the cable introduce low frequency (<< 2 Hz) into the DAS strain measurements. Low-cut filtering can remove these slow temperature trends from the DAS strain data. Converting the DAS strain data to strain rate effectively removes most of the temperature sensitivity.

Coupling of the fiber optic cable to the wellbore can be an issue in vertical wells using wireline or disposable fiber to acquire VSP data. In zones of the well where the cable is not well coupled to the wellbore, zigzag noise can be present. This noise is high amplitude and temporally persistent. One potential method to avoid this noise is by increasing the coupling aspects of the fiber to the wellbore. When that is not practical, there are several methodologies that are being used to filter out this noise post-acquisition.

A challenge for acquiring DAS VSP data in producing wells is the noise from production flow. This noise has been reported to be concentrated below 40 Hz. This is a significant portion of the seismic bandwidth, especially for deep wells where the center frequency of the reflected seismic wavelet may be only 20 Hz. Efforts to mitigate this noise will include smart digital filtering and high-fold acquisition where the random nature of the noise can be stacked out.

Chapter 9 Presurvey planning

The student who uses home made apparatus, which is always going wrong, often learns more than one who has the use of carefully adjusted instruments, to which he is apt to trust and which he dares not take to pieces.

– James Clerk Maxwell

I give you homework because I love you!

– Lucienne Osborn

9. Presurvey planning

Scientists and engineers always want more information, which drives us to acquire large complete surveys that will reveal the structural and stratigraphic details of the subsurface. Before actually purchasing and setting into motion the acquisition of these data sets, it is critical that the surveys are designed, modeled, and tested so that they will have a high probability of success. In this chapter, I will review some of the steps that can be taken to model and image data for surveys using active sources. I have developed an open source software package upon which you can practice and learn how the selection of the fiber optic acquisition parameters will affect the resulting migrated images.

9.1. ExploreDAS – an open source software package

I developed an open source, MATLAB graphical user interface software package for this course called ExploreDAS. The software allows you to quickly build a geologic structure, perform forward modeling to obtain seismic data, and create a migrated image with that data. The Appendix is the user's guide for the ExploreDAS software where I also provide a link for obtaining it. My objective in writing this software is to allow you to develop an intuitive understanding of DAS data sets and to compare them with equivalent geophone data.

9.2. Setting the objectives of the survey

Broadly speaking, there are three types of seismic surveys we acquire: structural imaging, stratigraphic imaging, and time-lapse imaging. Each of these have their own specific requirements. Structural imaging projects need to produce information about the overall gross shapes and features of the subsurface. As such, the S/N and frequency content requirements are typically not as stringent. Stratigraphic imaging projects need to produce the fine details of the subsurface formations so that the relationships and stratigraphic changes within each layer can be determined. This requires broadband frequency content of the seismic images. The objective of time-lapse projects is typically to look over a time interval for small changes in the structures or in the reflection amplitudes from formations that will reveal fluid flow, depletion, gas injection, fracturing, or subsidence. This requires a repeatable source with a broad frequency band, and a resulting survey with very high S/N.

9.3. *Illuminating the target formations*

In order to fully illuminate the target subsurface formations, the position of the sources and the location of the fiber must have the proper geometrical relationship with the subsurface structures. Like playing pocket billiards, the direction that the reflected seismic energy (billiard ball) takes must be aligned so that it hits the fiber cable (into one of the pockets) properly. With an easy-to-use modeling and migration software application, many different structural models can be evaluated to determine if the proposed acquisition geometry will be able to create useful images. Using a DAS data set to image the subsurface target of interest typically should be inferior to using 3C geophone data, so the acquisition geometry needs to be carefully studied to meet the imaging objectives.

9.4. *Designing surface seismic surveys*

Most surface seismic acquisition projects use vertical component geophones. Because the velocity of the near-surface is usually much slower than the deeper formations, seismic rays tend to bend upward toward the vertical as they come to the surface. Alternatively, seismic surveys can use 3C geophones on the surface and orientable seismic sources, such as shear-wave vibrators or tilted weight drops, which allows them to acquire 9C surveys. However, because we are interested in fiber optic cables, we want to understand the limitations of using a fiber cable on or near the surface.

Figure 9.1 shows a structural model (in blue) created to mimic a CO_2 sequestration study (Wilson et al., 2021). The red asterisks show the locations of 61 shots. The black line at the surface shows the location of the fiber optic cable. Forward modeling is performed for a P-wave source, with a Ricker wavelet at 50 Hz and a gauge length of 20 m. Figure 9.2a and 9.2b shows the geophone and DAS records, respectively. The direct arrivals have been omitted from the modeling so that only the reflected events are displayed. The red lines show the arrival times of the direct wave. It is obvious from the DAS record that the reflection events arriving normal to the fiber in the middle of the record have very low amplitudes. This is expected because these reflected events are arriving at the null of the P-wave response of the fiber. The question is how badly will this degrade the images of our target horizons.

Figure 9.3 shows migrated images of a portion of the model in Figure 9.1, using the modeled geophone and DAS data sets. From these two images, we can make some observations:

1. The DAS image does not reach as far laterally as the geophone image. This is due to the directivity of the fiber – the shots on the edges of the model are not contributing their zero-offset reflectivity.

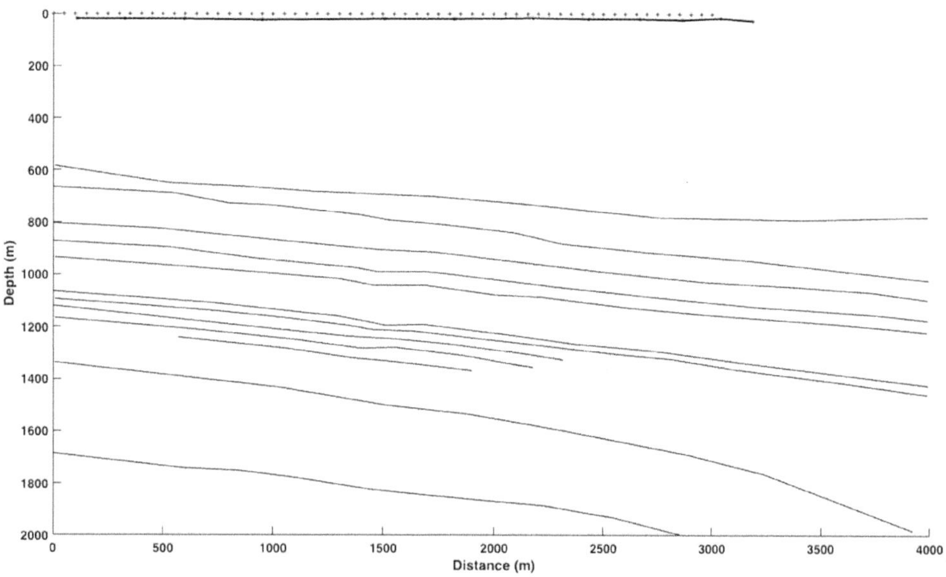

Figure 9.1. Example surface seismic acquisition geometry.

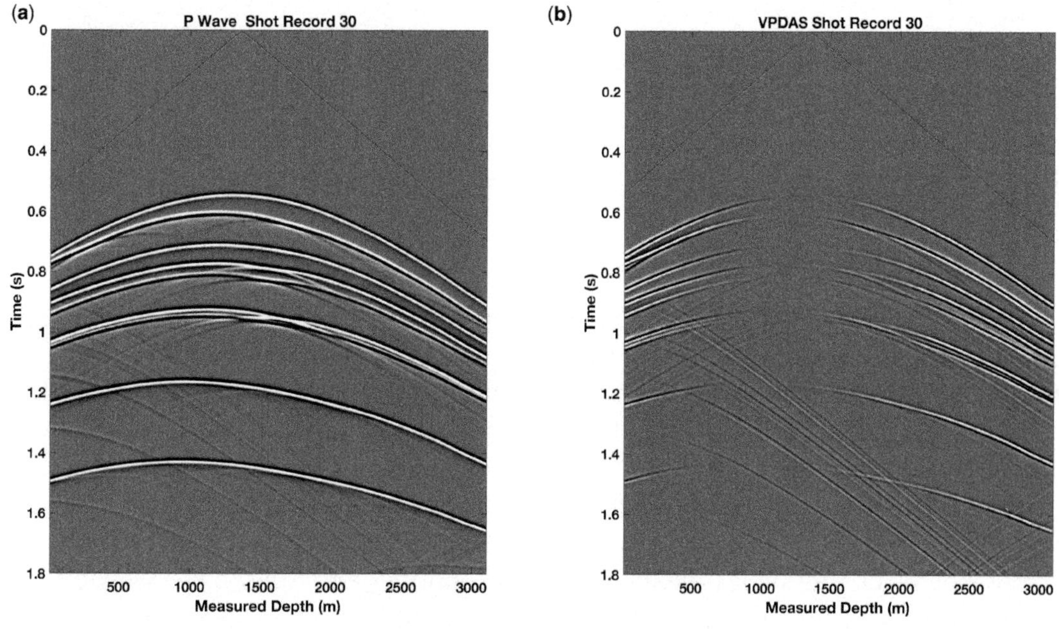

Figure 9.2. Synthetic shot record 30 (a) for geophone and (b) for DAS.

2. The wavelets between the DAS and geophone images are different. This is expected and can be corrected using the equations in Chapter 3.
3. The migration sweep artifacts are stronger in the DAS image than in the geophone image. Again, this is due to the directivity of the fiber.

Additional tests can be run using different source wavelet types (Ricker or Klauder), source bandwidths, gauge lengths, etc. These will build your understanding and intuition about what factors are important to meet your imaging objectives.

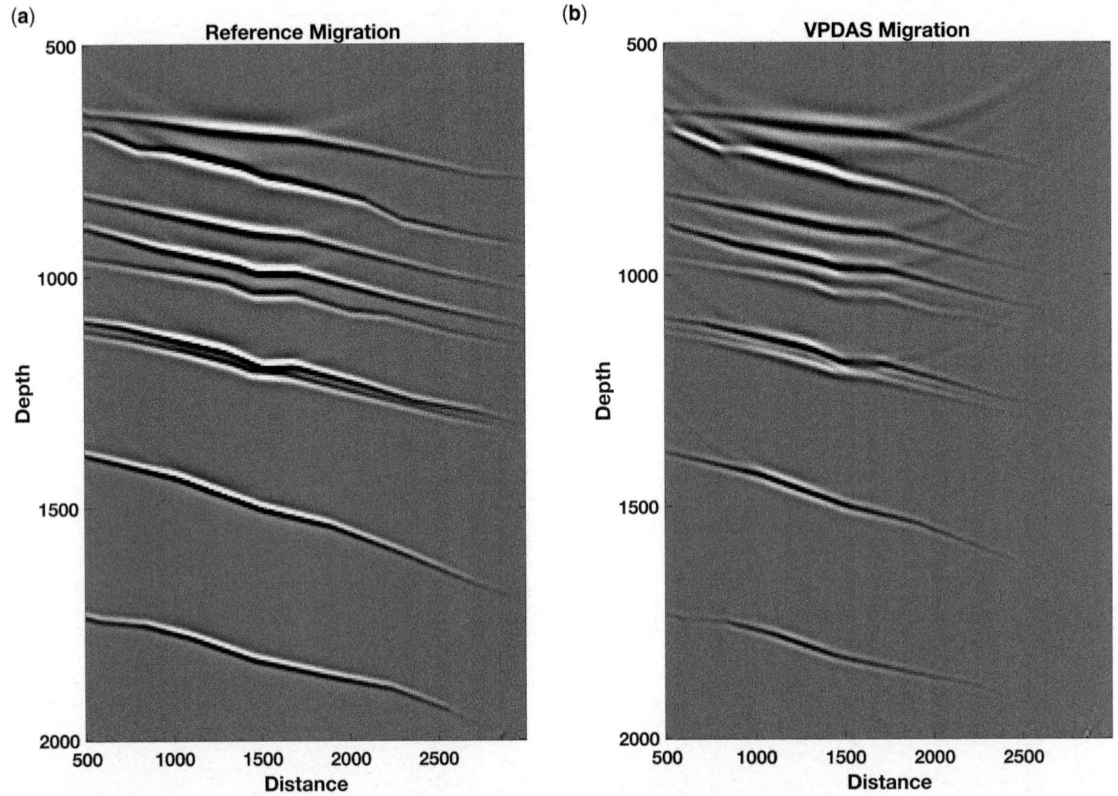

Figure 9.3. Migrated images for (a) geophone and (b) DAS strain rate input data.

9.5. Designing VSP surveys

In contrast to surface seismic acquisition, most VSPs use 3C geophones. Having access to all three components of the wavefield means that we can perform Alford rotation to point them in any direction we might need. However, for VSPs acquired with fiber, we will be limited to only the "vertical" component which is aligned along the axial direction of the fiber.

Figure 9.4 shows a model that was created to represent a complex subsurface structure (Willis et al., 2020, 2021). As previously, the red asterisks show the location of 71 shots, the black line shows the locations of geophones and fiber in the well, and the blue lines indicate the layer boundaries. Seismic data have been created using a P-wave source with a Ricker wavelet, centered at a frequency of 50 Hz and a gauge length of 25 m. Figure 9.5a shows the geophone data and Figure 9.5b shows the corresponding DAS strain rate data. The records do not include the direct P-wave and there is a small mute around the direct arrival time window.

As for the surface seismic example, there are missing events on the DAS strain rate data when the events arrive normal to the fiber. Figure 9.6 shows the migrated images for the geophone data (left) and the DAS strain rate data (right). Overall the images are very similar, but as expected the wavelets are different. In addition, some of the deeper dipping events are missing on the DAS image (for example, at distance = 4000 m and depth = 2000 m).

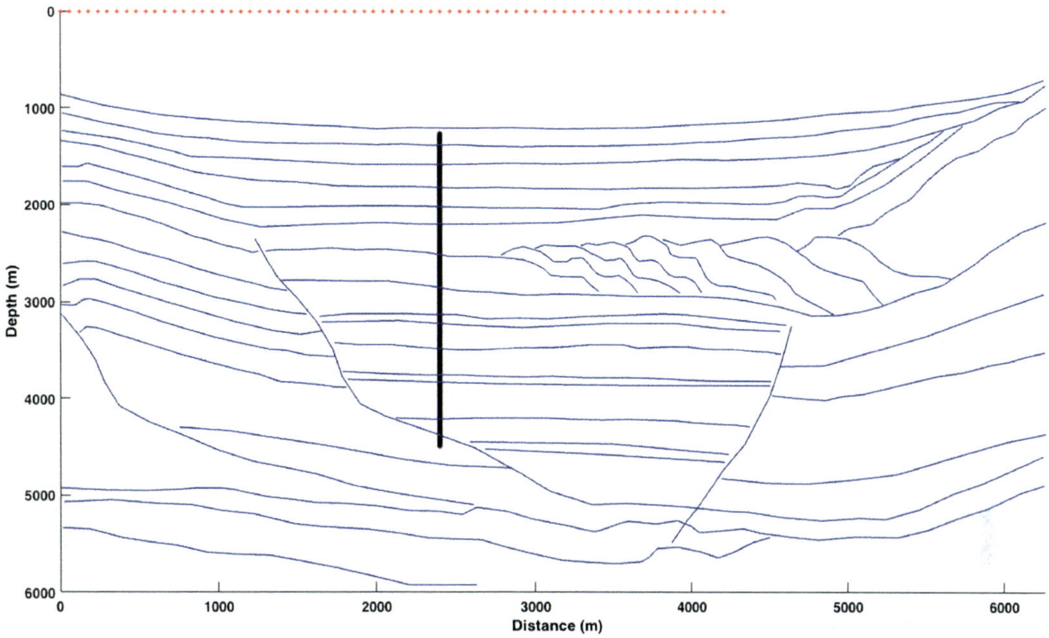

Figure 9.4. Structural model showing sources (red asterisks), fiber in the well (black line), and layer boundaries (blue lines). (Modified from Willis et al., 2020, Figure 1.)

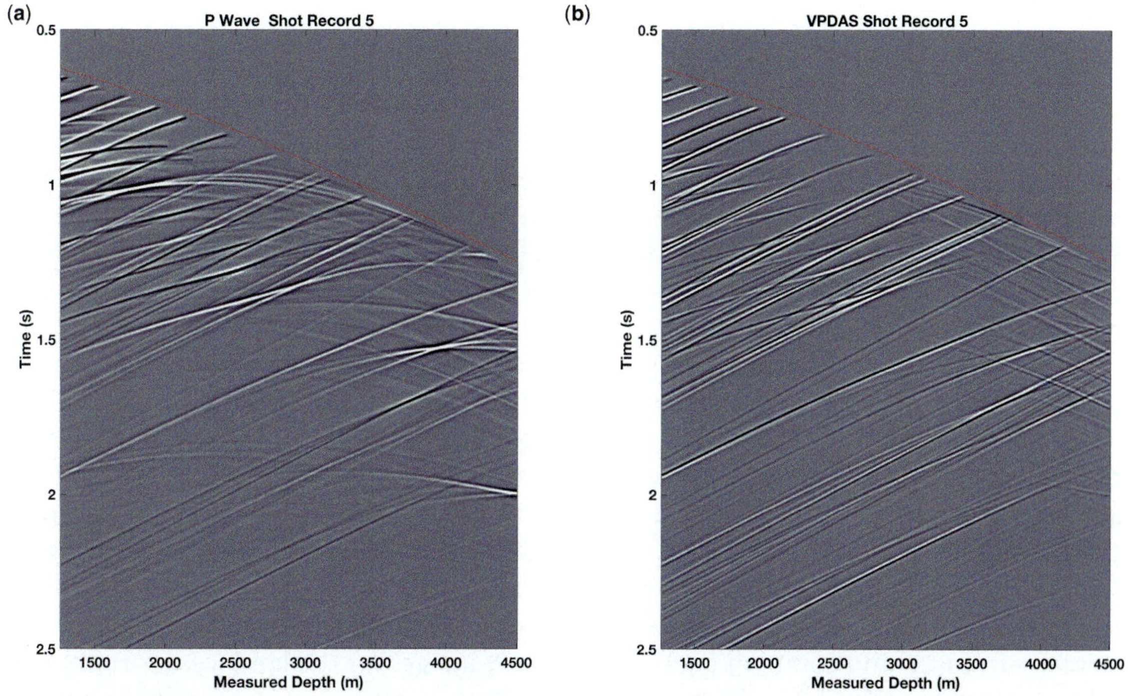

Figure 9.5. Synthetic VSP records for (a) geophone and (b) DAS strain rate. The direct arrival time is shown by the red line.

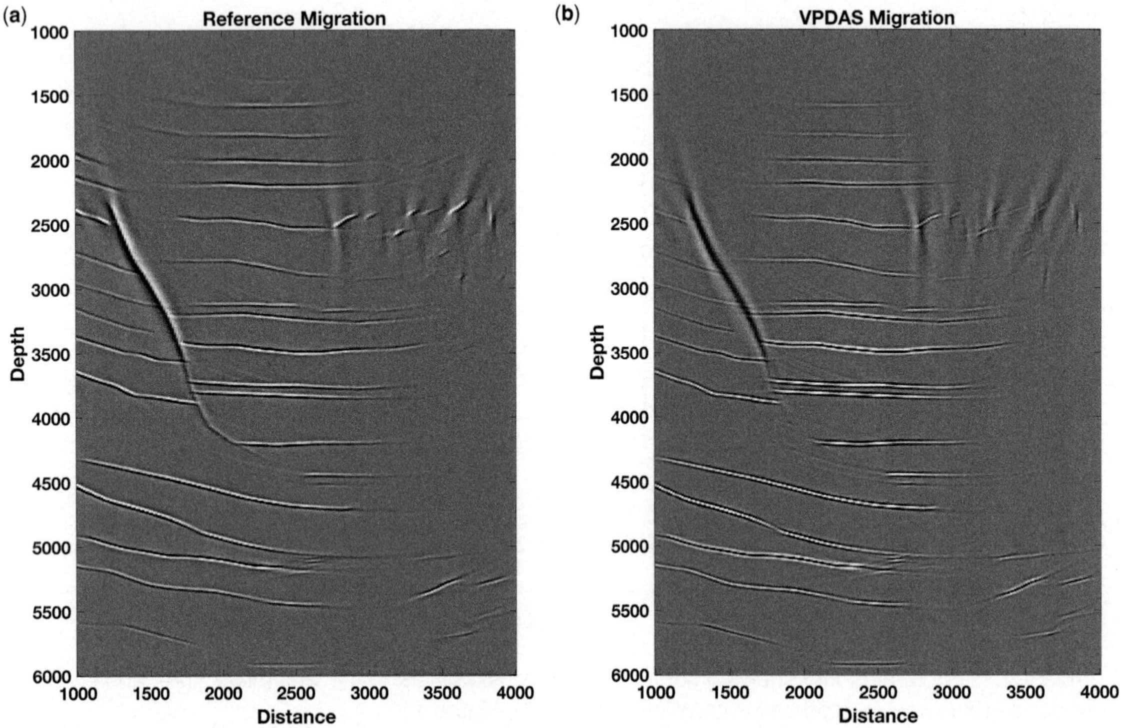

Figure 9.6. Migrated VSP images from (a) geophones and (b) DAS strain rate data.

9.6. Studying the effect of optical noise

Noise is an ever-present part of recording any type of measurement. I have described several sources of optical-related noise on strain (rate) data in Chapter 8. With the ExploreDAS software, you can study the effect of optical noise on the raw modeled seismic records and also on the migrated images. There does not seem to be another such tool available to easily quantify the effect of this noise. I have used this software as the basis to create synthetic data and images for papers to teach the need for solutions to provide low noise DAS data sets (Willis et al., 2020; Wilson et al., 2021).

Figure 9.7 shows a compilation of images demonstrating the effect two different levels of optical noise on a synthetic data set. The top row of the image shows a magnification of the subsurface model (left), the migrated image from DAS strain rate records containing the expected typical levels of optical noise (S/N = 20 dB) for land acquisition (center), and the migrated image adding the expected level of optical noise (SNR = –1.4 dB) for a straightforward, simplistic DAS acquisition system for subsea data (right). From these images, it is clear that there is a significant loss in image quality for the subsea case. The bottom row shows shot records for the land (left) and subsea (center-left) cases, and common depth point (CDP) images for the land (center-right) and subsea (right) cases. It is clear from the subsea CDP image that any angle gather work on the subsea migrated data would be extremely challenging.

As I discuss in Section 5.4, there are ways to improve the subsea data quality based upon substituting or adding additional hardware, such as ultra-low loss fiber in the umbilicals and a seafloor remote circulator. However, without studying the impact of

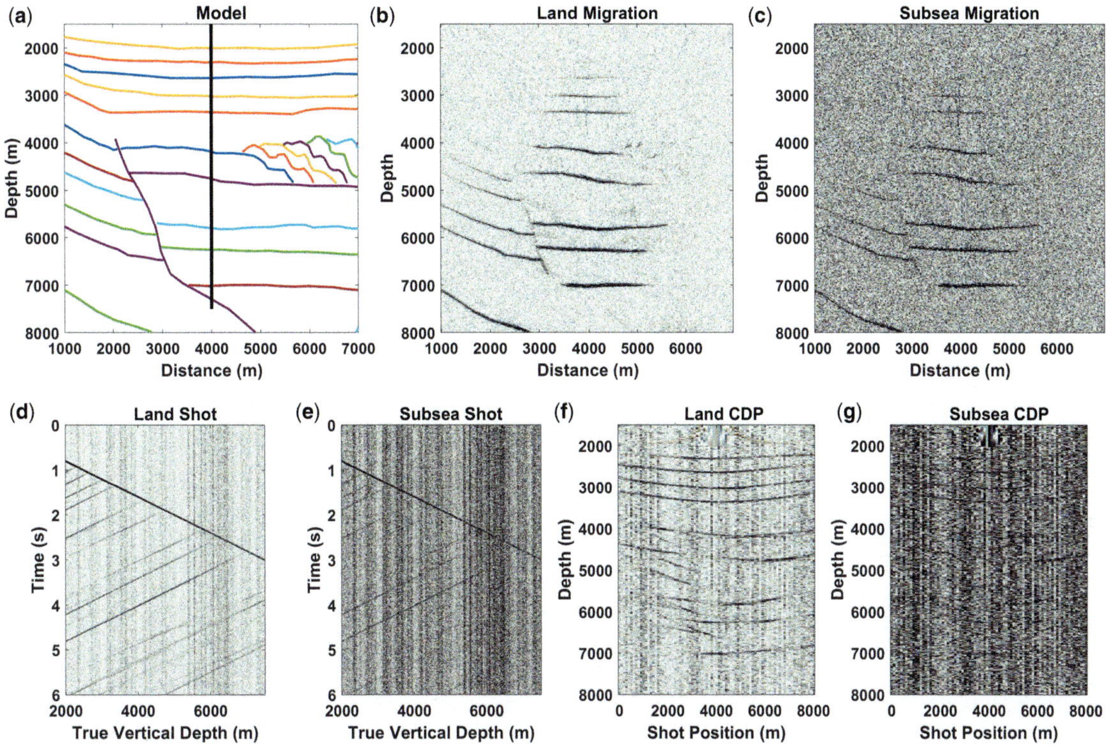

Figure 9.7. Effects of optical noise on DAS strain rate data. (a) Structural model. (b) Migrated image for typical land level of noise. (c) Migrated image for expected subsea level of noise. (d) Shot record for land case. (e) Shot record for subsea case. (f) Land migrated CDP. (g) Subsea migrated CDP. (From Willis et al., 2020, Figure 1.)

the S/N of the shot records on the final image, one is flying blindly into data acquisition, not knowing what the needed S/N should be. I hope that you will find the ExploreDAS software useful in specifying your S/N requirements.

9.7. Summary

Each seismic survey will have different objectives based upon the structural, stratigraphic, and time-lapse goal of the project. Before committing to any seismic acquisition design, it is imperative to perform presurvey modeling and migration tests. The acquisition geometry must be designed so that the seismic energy will illuminate the target formations under consideration. Using fiber optic cables for acquiring strain (rate) data is a new part of seismic survey design as most tools have been designed for geophone surveys. By using the ExploreDAS software, you can quickly design any arbitrary 2D survey geometry, perform forward modeling of strain rate data, and create migrate images. You can study the effects of the gauge length, source wavelet and frequency band, and optical noise on the survey products in order to build an understanding and intuition for your future projects.

Chapter 10 Field deployment issues

Farming looks mighty easy when your plow is a pencil and you're a thousand miles from the corn field.

– Dwight D. Eisenhower

10. Field deployment issues

In many ways, fiber optic seismic data acquisition is similar to conventional geophone data acquisition. I won't cover these common aspects but will give an overview of some of the unique aspects in this chapter.

10.1. Fiber coupling issues

One of the most important factors for the success of a DAS seismic data acquisition project is to ensure good coupling of the fiber to the earth. There are different aspects to this depending upon whether the fiber will be on the surface or in a borehole.

10.1.1. Surface fiber

Simply laying a fiber optic cable on the surface of the earth is perhaps the most economical deployment method. However, an issue with laying any type of cabling on the surface of the ground is tripping hazards. As part of a surface seismic geophone acquisition project at one of my previous employers, the land owner's prize horse tripped over the geophone cabling and broke its leg. This was a tragedy for the horse and the owner, but also was expensive for my employer. So, on all counts, care must be taken to avoid such issues.

As I mention in Chapter 2, two approaches to achieve better coupling are pushing the cable into damp soil (Martin et al., 2017) or, if the temperature is sub-freezing, pouring water over it to freeze the cable to the ground surface (Yang, 2021). Mjehovich and Jin (2021) study the coupling efficiency when laying fiber on the grass, snow, and concrete on the Colorado School of Mines campus. They also compare it against trenched-in fiber. From all of the deployment methods, they obtain good data. In the snow case, however, they find better coupling of the surface fiber when it is pressed into the snow by walking over it rather than simply laying it on the snow. (As I mention in Section 10.2, care must be taken not to damage the fiber when handling it, whether on the surface or when burying it.) The trenched-in fiber has the best response but obviously costs much more to implement. The practical aspects for trenching fiber need to consider crossing obstacles such as roads, ponds, and nature preserves.

Existing dark fiber inside telecommunication conduits has been used by many studies to alleviate the need to deploy special-purpose fiber (Martin et al., 2017; Jakkampudi et al., 2020). The practical issue is the inability to place the fiber where it is needed and the likely tortuous path it takes to supply the telecommunication fibers to the appropriate infrastructure and buildings.

10.1.2. Borehole fiber

If fiber or geophones are deployed into a cased well which is not cemented properly, the resulting seismic data will contain artifacts in the zones with poor cement. Good cement is a requirement for obtaining good DAS data. It is quite valuable to have a cement bond log run to check on the condition of the cement before contemplating acquiring DAS if there is any doubt about the quality of the cementing. This will help to ascertain where the artifacts may occur and if it is still worth acquiring the DAS data. For open hole logging, this is not a concern.

As mentioned in Chapter 8, if wireline or disposable fiber is to be deployed into a vertical well, there is a good chance that several zones will have zigzag noise from the cable not being in good contact with the wellbore. One mitigation strategy is to adjust the slack on the wireline as is done routinely for borehole geophone acquisition projects. Perhaps even a systematic slacking can be done (Frignet and Hartog, 2014). Another possible approach is to acquire data with the cable pulled up into several maximum depth positions as some of the zigzag noise will move to different depths (Willis et al., 2019b). This would require repeated source excitations, however.

If the zigzag noise cannot be avoided during acquisition, there are many signal processing methods that can remove the noise (Cai et al., 2016; Yu et al., 2018; Chen et al., 2020; Willis et al., 2020; Hou et al., 2021; Martuganova et al., 2021).

10.2. Fiber handling

Field technicians need to be fully trained to handle fiber because fiber cables don't behave like conventional electrical cables. Optical connections need to be made extremely carefully, with the appropriate equipment and skill in order to eliminate any unnecessary losses. These connections take place at multiple locations along the fiber's path into the wellbore, for example, at the UI, at the surface cable connection to the take out box, and at the well head.

The radius of curvature of the fiber optic cable must be maximized to maintain the optical integrity (Blount, 2015). Bending the cable tighter than the manufacture's recommendation will allow light to leak out of the fiber's core and large optical losses will ensue. This is generally called macrobending, where curvatures on the order of a few centimeters will induce losses (Jay, 2010). Another aspect is that pulling and bending the cable around a corner, such as around a pulley, may damage the cable itself. There are standard methods for measurement of the losses from macrobending (IEC 60793-1-47 Measurement methods and test procedures – macrobending loss; IEC 60793-2-50 Product specifications – sectional specification for class B single mode fibres) in the laboratory.

Microbending is a term used to categorize the signal losses from defects in the waveguide effectiveness of the fiber from either the manufacturing process issues or mechanic stress such as pressure, tension, and twist of the fiber (Blount et al., 2015). Jay (2010) describes how microbending can occur when the glass fiber is pressed against an irregular surface. Thus, it is critical to keep the fiber from being stressed during the handling, installation, and maintenance process, for example, from tie-down clamps.

While the recommendations of the manufacturer of the fiber optic cable must be followed, a general guideline given by the Fiber Optic Association (www.thefoa.org/tech/ref/install/bend_radius.html) is that the radius of curvature for the cable during installation while it is under tension, $R_{tension}$, and after installation, $R_{installed}$, should be given by

$$R_{tension} \geq 20\ d, \tag{10.1}$$

$$R_{installed} \geq 10\ d, \tag{10.2}$$

where d is the diameter of the fiber optic cable.

10.3. Fiber health

Before any seismic survey is taken using a fiber optic cable, it is important to ascertain its health – is the fiber still able to carry an optical signal and hasn't been damaged by either improper handling or hydrogen darkening. A tool used to make this determination is an optical time-domain reflectometer (OTDR), which is described in detail in Chapter 3 of Hartog (2017). By connecting this device to the fiber optic cable at the surface take out box, the health of the deployed fiber can be measured ahead of any contemplated survey being performed. Typically, more than one optical fiber is in the cable, so the measurement should be made on all of the fibers to determine the best fiber to use.

Figure 10.1 shows an illustration of the read-out of an OTDR measurement. The horizonal axis is the distance along the fiber path converted from the transit time of the optical pulse using equations 5.1 and 5.2. The vertical axis is the reflectance of the

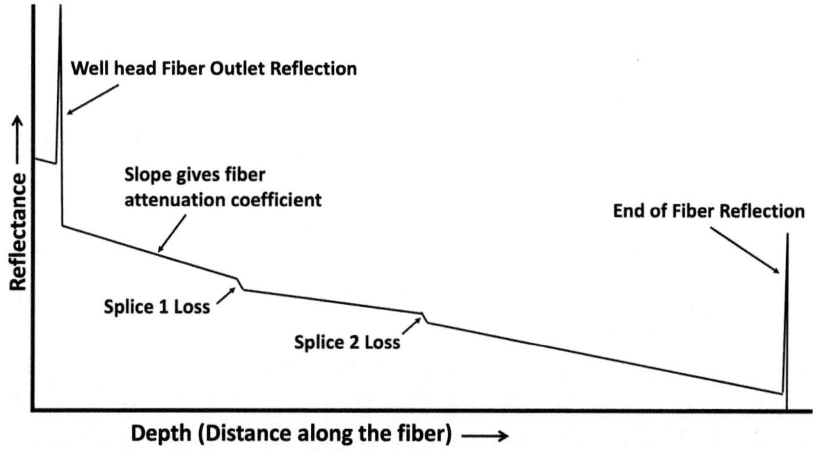

Figure 10.1. Illustration of an OTDR reading showing large losses at the well head and end of fiber, and small losses at the two splice points. The slope of the curve is a measurement of the attenuation of the fiber optic cable.

backscattered light in the fiber, displayed in dB, which has been scaled to account for the two-way travel path (https://www.thefoa.org/tech/ref/testing/OTDR/OTDR.html).

The event labeled "Well head fiber outlet reflection" is the optical reflection that is coming from the fiber optic connection at the well head. It shows two features – a large reflected wavelet, which comes from the pulse of light emitted by the OTDR, and to the right of the wavelet there is a drop in the amplitude of the backscattered energy. This indicates that the connection has a large reflection coefficient and an associated large loss in the transmission coefficient.

The event labeled "End of fiber reflection" is the reflection coming from the end of the deployed fiber (whether that is in the well or on the surface). It has properties similar to the well head reflection, however, the reflection coefficient is almost unity. The fiber can be terminated to reduce this reflection before it is deployed, as the reflection from the end of the fiber can harm an IU.

There are two events labeled "splice loss 1" and "splice loss 2." In this example, the fiber optic cable has been spliced at these two locations which causes optical losses, but they don't usually cause reflections as their impedance contrast is small.

As we've seen, the events on the OTDR display can indicate the location of optical connections and splices made in the cable. But it also displays the optical attenuation of the fiber itself. Between the events on the displays, we see three different sloping lines. The measured slope of these lines provides the attenuation of the fiber in each of these sections. It is clear that each portion of the cable is exhibiting a different slope and thus a different attenuation with distance. Of key concern is whether any of the sections of fiber have very high losses which would make the fiber unusable for a DAS survey. The OTDR machines have markers on the screen display that let you determine the losses across connections and also determine the slopes of the losses in the fiber cable. The attenuation loss in dB/km, associated with the measured slope, should be checked against the specification sheet of the fiber to make sure that the fiber has not been damaged.

10.4. Depth calibration

It is extremely important to be able to match the channel numbers of the DAS data acquired with the corresponding distance or depth of the measurement. There are several approaches to perform this depthing operation.

10.4.1. Depth calibration using nominal values

As we've seen, the strain (rate) measurement is made by pulsing light into a fiber optic cable and waiting for the backscattered light to return to the IU. We convert the traveltime of the pulse for each measurement to distance or depth using equations 5.1 and 5.2. This is done using an estimate for the velocity of light in the fiber, typically from the specifications given by the manufacturer of the fiber optic cable.

We also have seen, in Figure 5.4, that the manner in which the fiber is wound into the cable can greatly change the amount of fiber present per unit distance of the cable. Straight fiber, without any additional winding of the fiber, is typically over-stuffed by 10% into the cable to avoid it being damaged by tension on the cable. If that factor is not taken into consideration when converting the channel locations to depth, they will

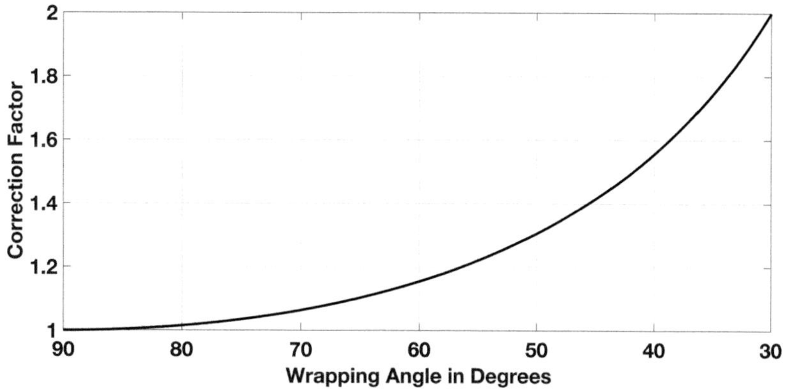

Figure 10.2. Correction factor required if the wrapping angle of the optical fiber is not accounted for in the depth (distance) conversion.

likely be 10% too deep even if the correct velocity is used for the light in the fiber. For helically wound fiber, the depth errors accumulate must faster.

Figure 10.2 shows the correction factor required if the wrapping angle is not taken into consideration when converting the channels to distance. The horizontal axis shows the wrapping angle and the vertical axis shows the theoretical factor to apply to the cable length to obtain the corresponding fiber length. For a perfectly straight fiber, 90°, no scaling is required. For a 60° wrapping angle, the scaling factor would be approximately 1.1547. For a 30° angle, the scaling factor would be approximately 2, meaning the fiber is actually twice as long as the cable.

I've shown there are ways to convert the traveltime propagating in the fiber to the corresponding depth (distance) of each channel using nominal information – the specified index of refraction of the fiber and its winding angle. This gets us close, but is not necessarily as accurate as we need.

10.4.2. Depth calibration using reference points

An important method for depth calibration is to use known control points, which are places along the fiber cable that can be assigned to known positions. Tap tests can be performed wherein someone taps on the fiber or near the fiber at a known location (Ellmauthaler et al., 2016a). Then, the measured DAS sound of the tapping is identified on the waterfall plot and the known location of that channel can be registered. A waterfall plot is a display that is created in the field which shows the amplitude of each channel of the fiber as a function of time. Figure 10.3 shows a waterfall plot of the DAS signal where the vertical axis is the channels along the fiber and the horizontal axis is time. The color of each sample is related to the amplitude of each sample in the DAS record. The arrows point out the location along the fiber and the timing of two taps at the well head. Then, this location can be used as a known reference point on the fiber. For fiber optic cables on the surface, in trenches, or in telecommunication conduits, it is straightforward to perform as many tap tests as needed to calibrate the location of the fiber. For fiber in a well, the well head is the most easily accessible tap test location.

Other known location points on the fiber are likely to be the end of the fiber and recorded locations where the fiber has been spliced. For fiber in a well, there may be splices made for pressure sensors that could be used as control points.

10.4.3. Depth calibration using cable pulls up

For DAS VSPs using wireline-deployed fiber in a well, there is another approach that can be used for mapping the channel number to depth in the well (Ellmauthaler et al., 2018). First, the cable is deployed to depth and the survey is acquired. Before pulling out of the hole, the cable is moved up the well, preferably 600 m or more, and at least one shot location is repeated. The depth of the lowest channel of the cable is determined for both acquisitions, using standard wireline depth control procedures. Then, these two locations can be used to depth the rest of the channels.

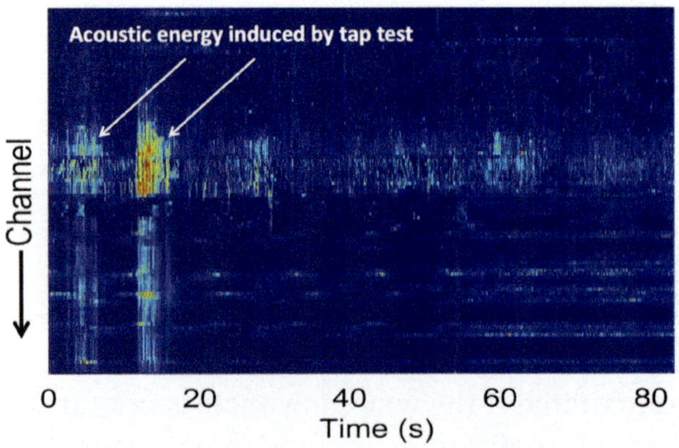

Figure 10.3. Waterfall plot of the DAS signal showing two taps on the well head to calibrate the fiber. (Modified from Ellmauthaler et al., 2016a.)

Figure 10.4 illustrates acquiring DAS VSP records with the wireline

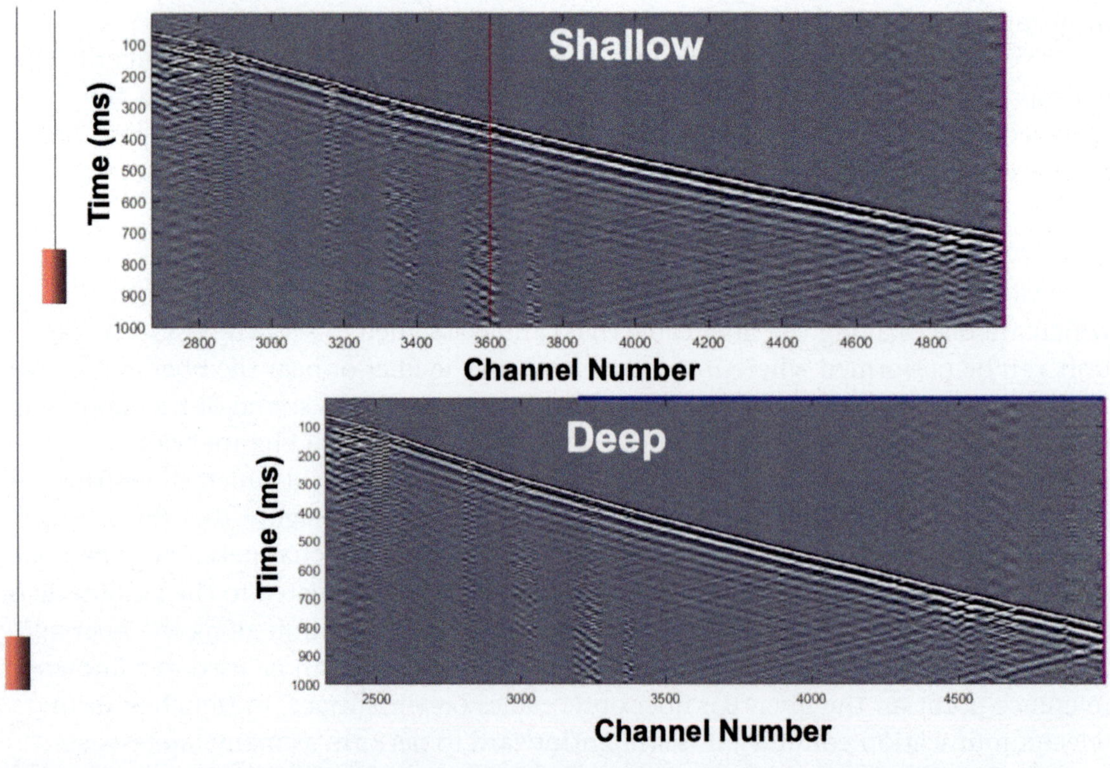

Figure 10.4. Illustration showing acquisition of a "Deep" DAS record with the wireline near the bottom of the well (bottom) and another "Shallow" record with the wireline pulled up 350 m in the well (top).

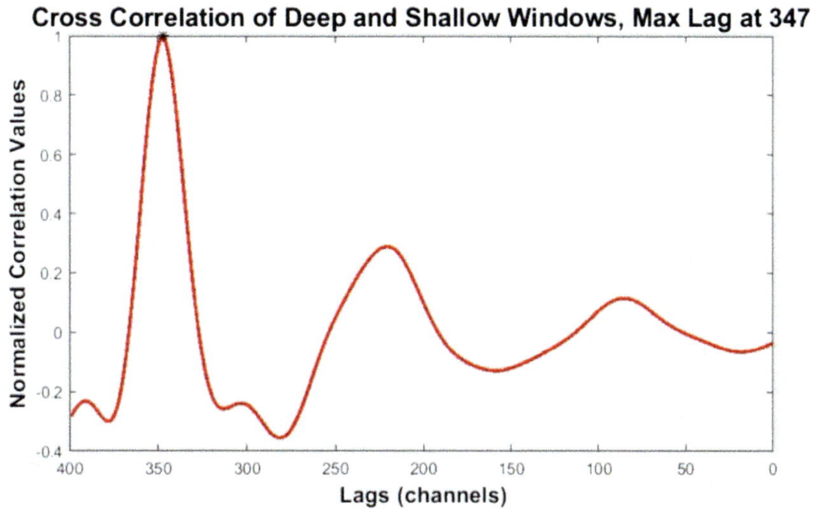

Figure 10.5. The cross-correlation function comparing the Deep and Shallow records shown in Figure 10.4.

fiber at two different maximum depth locations. The bottom record labeled "Deep" has been recorded with the wireline near the bottom of the well. The upper record, labeled "Shallow," has been recorded with the wireline pulled up 350 m. The two records shown have been approximately aligned in depth in the figure for ease of comparison. Because this is wireline data, the channel numbers are fixed with respect to the cable, so the depthing must be performed to match the depth of all channels with respect to the known depth of the lowest channel in the well.

Figure 10.5 shows the cross-correlation function comparing the deep and shallow DAS records. The maximum peak at 347 traces shows the best alignment between the two records. Because the depths are known now at the bottom of each of the records, the depth of each channel can be computed accurately. Figure 10.6 shows the two records merged together which confirms the depthing is accurate.

10.4.4. Depth calibration using tube waves

Tube waves propagate along the borehole at a predictable velocity (Paillet and White, 1982). Tube waves are generated in the VSP records especially when the source offset to the well head is small. The moveout of the tube wave can help to calibrate the depth of the channels based upon an estimate of the tube wave velocity (Verliac et al., 2015). Any changes in the casing or open hole diameter size of the well will create reflected tube waves which can be used for depth calibration if the depths of hole size changes are known. In a similar way, water hammer waves can be created by valves opening and closing in the fluid lines near the well head which can be used for depthing (Granville and Willis, 2021).

10.4.5. Depth calibration using inversion of arrival times

A formalized inversion of the arrival times at each channel from multiple shot locations can be used to invert for the channel depths and subsurface velocity structure (Padhi et al., 2021). The arrival times for each channel and all shots are picked. The

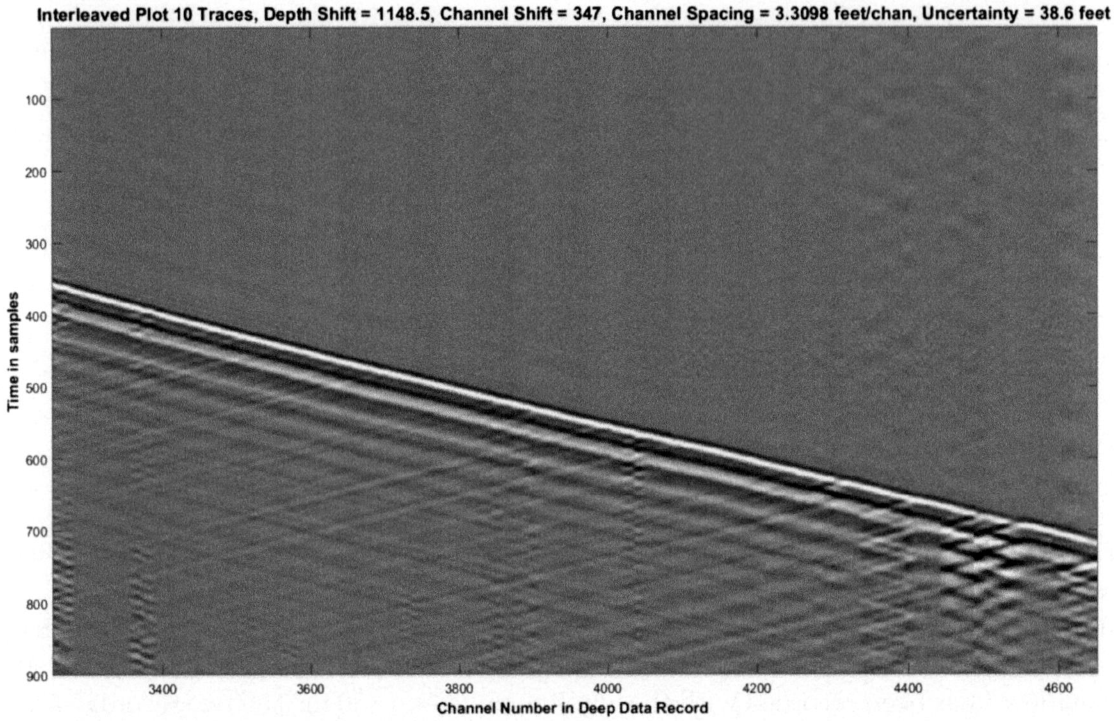

Figure 10.6. Merge of the Deep and Shallow records shown in Figure 10.4 after depthing.

depth of each channel is known approximately at the start but can be updated with the inversion algorithm given constraint criteria. The set of arrival times are fed into a globalized inversion process which estimates both the location of the channels in depth and the velocity model. This procedure would likely be done much later after acquisition during processing of the data.

10.4.6. Depth calibration using migration

Oloffson and Martinez (2017) suggest many possible sources for errors in the depthing of each channel. They suggest that the DAS VSP data are sorted into common receiver (channel) gathers which then are depth-migrated separately. Then, the common receiver migrated images can be shifted and adjusted so that the depthing of each channel is resolved.

10.5. Data processing and transmission

One of the first issues in the field will be how to quality control the DAS data being acquired. If the recording is being triggered by active shots, then this will be easy because individual shot records can be viewed and their S/N can be determined (Wilson et al., 2020b). Alternatively, if the DAS data will be recorded continuously, then one approach for field quality control is to observe the data using the waterfall plot (as described in Section 10.3.2). Typically, continuously recorded data is transmitted or transferred in bulk back to the processing center and separated into shot records for processing and quality control.

Table 10.1. Example raw data DAS VSP set sizes for 2D and 3D surveys.

	2D Survey	3D Survey
Number of shots	1,000	60,000
Number of wells with fiber	1	4
Fiber channels per well	5000	5000
Record length (sec)	20	10
Data volume	80 GB	19.2 TB

The next decision that needs to be made is how the DAS data will be stored in the field and then transferred to the client and/or data center for processing. The amount of data is much greater than for geophone data sets. Table 10.1 shows that, for an example 2D DAS VSP data survey, the raw uncorrelated data is fairly manageable at 80 GB. However, when looking at large marine 3D surveys, the data sets can easily exceed 20 TB, which is time consuming and expensive to transfer via satellite. For continuously recorded acquisition, the data sets can become much larger.

Redundant data acquisition systems need to be employed in the field so that, in the event of disk failures on one system, there is always another copy. Dual-RAID (redundant array of independent disks) disk systems should be employed which balance the need for fast disk writes using striping with the redundancy of parity and mirroring.

If a LAN network connection is available at the recording station, data can be streamed continuously back to both the client and the data processing center. In remote locations, data streaming by satellite is potentially an option. If neither of these options is available, then physically transporting the hard drives from the field location back to the processing center is the best option. Because these hard drives will have come from RAID-systems, it imperative that the drives are labeled properly so that they can be installed in the destination RAID computer systems.

10.6. Source issues

If the recording is being triggered by active sources, then radio communication is imperative to connect the source controller(s) to the triggering system of the IU. If radio communication is lost for a time window during acquisition, then those corresponding shots will not be recorded. Scouting the site in advance is important to ensure that radio communication is possible to and from all locations of the survey and will not be blocked by infrastructure or landscape topography.

If continuous recording is the selected option, then a reliable GPS time system needs to be employed. This includes both the GPS receiver and a good antenna. Prior to the availability of GPS systems, frequently the internal computer clock values have been used to time-stamp files, but this poses problems with time drift, knowing what is the local time zone, and coordinating between different acquisition computers. A colleague once suggested to me that we could record a local country western radio station in one data channel as a cheap way to keep track of time. For modern applications, neither of these methods is sufficiently accurate. When the continuous data is recorded, the GPS time-stamps on the data will be used along with a tally of the GPS times for each shot to cut the continuous data into shot records for active source surveys.

Fiber optic cables have less sensitivity than geophones. Geophones typically have a much lower noise floor and thus can use smaller seismic sources. So, optimizing the active source effort is important for acquiring good DAS data. We can use the same decades-old options we've had for geophones to improve the DAS data S/N by increasing the source size and duration. Multiple vibrator trucks can sweep together to increase the source power. The vibrator sweep length time can be increased and the bandwidth tuned. Permanently installed orbital vibrators can sweep for minutes in time and substantially vary their frequency content to optimize data quality (Freifeld et al., 2016; Spackman, 2019). The charge size can be increased for explosive sources. Repeat firings of the source at the same location can increase the S/N by allowing stacking of the recorded data. In the marine environment, the number and size of air guns can be varied to optimize the desired S/N.

10.7. Safety in the field

It is critical to follow the best safety practices in the field. My father, David Willis, was the Party Chief and Computer on a land seismic crew for Keystone Oil Company back in the 1950s. They were using explosives for the seismic sources on surveys in Louisiana, Texas, and California. One of their safety routines was to impress upon the crew weekly how dangerous dynamite is and how important it is to handle it properly. To implement this practice, they would strap a small amount of it to a fence and then detonate it. On this one occasion, they had a new Shooter and my father told him to "strap a 1/8th of a stick of dynamite on the fence." At the designated time, the crew all gathered around, at a safe distance, and the lesson started. The explosion nearly knocked over their cars and trucks while showering all of them with debris. Instead of a 1/8th of a stick, the Shooter had put 8 sticks of dynamite on the fence. They are all lucky to have lived to tell the tale.

The lesson I want to convey is this: even when we think we have good safety practices in place, good communication and attention to the details are critical for everyone's safety.

10.8. Summary

I've presented some field strategies for obtaining good-quality DAS data. It is important to have the sensing fiber in contact with the earth in a way that provides good coupling with the seismic waves we want to measure. Fiber optic cables must be handled very differently from electrical cables. Sharp bends, tension, and pressure must be prevented to avoid damaging the cable and loss of signal quality. Before a survey is acquired, an OTDR survey of the existing fiber optic cable should be performed to determine if the fiber has good health and will sustain acquisition of DAS data without significant losses. The DAS data need to have the location of each channel determined using one of several methods. Plans need to be in place for recording, safeguarding, and transferring large quantities of DAS data which may exceed 20 TB. The source effort will likely need to be increased over geophone acquisition to ensure the acquired data quality.

To use data from DAS technology, you are not required to be a fiber optic expert. However, you need to be confident that it is giving you useful information. In this book and course, I have endeavored to share with you the high-level basic physics about how

DAS works using the backscattered light in an optical fiber with a delay gauge coil to extract the strain characteristics of a seismic wavefield. I've provided a set of equations that will allow you to convert DAS strain and strain rate data to and from the geophone response which enables the direct comparison of conventional seismic data with DAS data. I've defined the two basic types of optical noise, common mode noise and fading, and described mitigation strategies to address them. I've provided a comprehensive analysis for the unique characteristics of the DAS amplitude response describing the effect of the gauge length, apparent seismic velocity, and angle of incidence. I've also described how to select the gauge length to obtain optimal seismic data quality for the application you might be considering. As an additional bonus, I've provided an open source MATLAB software package that will allow you to design a DAS survey, create synthetic DAS data, and build subsurface images from them.

With all of these resources in your toolbox, it is my desire that you are now better equipped to embark on seismic acquisition and processing projects using DAS technology.

Appendix ExploreDAS software documentation

Open source is about collaborating; not competing.

– *Kelsey Hightower*

In open source, we feel strongly that to really do something well, you have to get a lot of people involved.

– *Linus Torvalds*

A. ExploreDAS – an open-source software package

Software applications are a key part of a scientist's and engineer's toolbox allowing them to stand on the shoulders of those who have gone ahead of them and apply their learnings to new problems.

I developed this open-source, MATLAB graphical user interface software package, called ExploreDAS, for this course to help you build a more intuitive understanding of the strengths and potential weakness of DAS seismic data acquisition and processing. The software allows you to quickly build a geologic structure, perform forward modeling to obtain seismic data, and create a migrated image with that data. It is available for download from Willis, 2022e.

My objective in writing this software is to allow you to develop an intuitive understanding of DAS data sets and to compare them with equivalent geophone data. Some of its characteristics are:

1. Creating structures, fiber geometry, and shot locations is extremely easy.
2. Layers with any arbitrary geometry can be formed.
3. It has two velocity options – constant or v(z).
4. Each layer has the same reflection coefficient.
5. It creates geophone data and the corresponding strain rate data.
6. Only "straight" fiber is modeled with the software (i.e., no helically wound fiber).
7. The gauge length can be changed as desired.
8. The P- and S-waves are modeled separately.
9. It uses a very simple Kirchhoff migration algorithm to form the migrated images.

Several disclaimers to note:

1. I don't claim that the amplitudes are totally accurate but were created to be sufficient for teaching purposes.

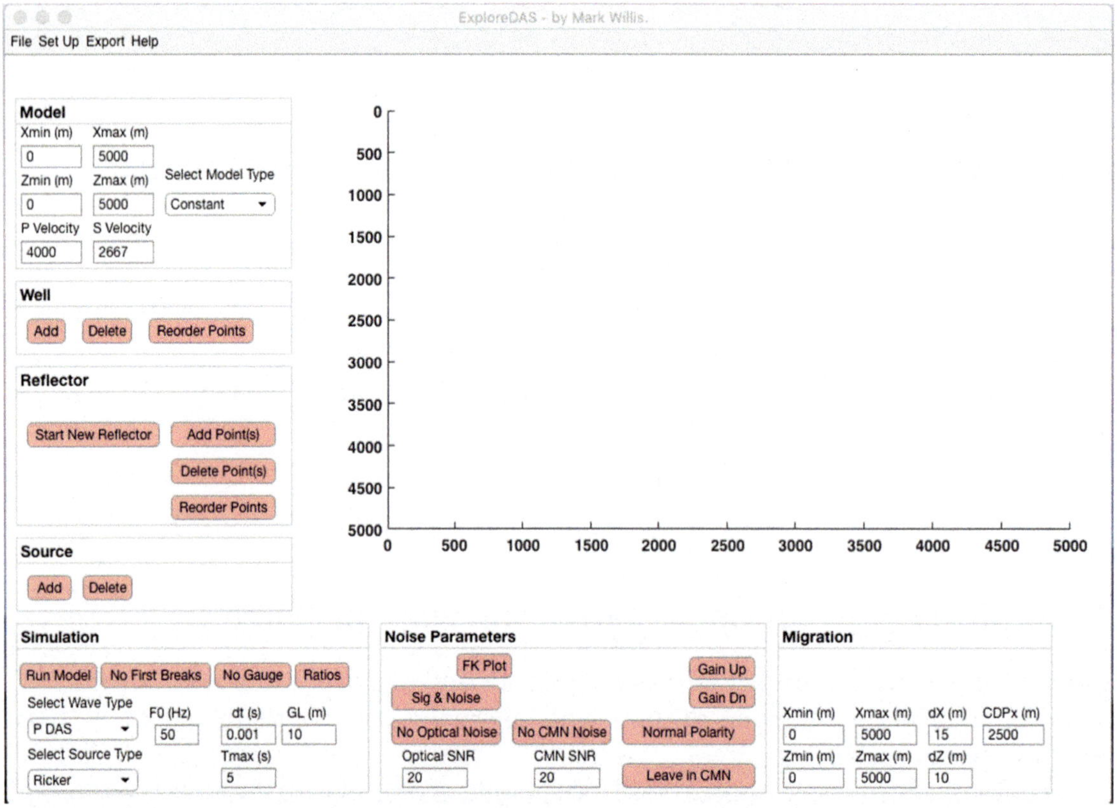

Figure A.1. The opening user interface for ExploreDAS.

2. Many better modeling and migration algorithms are available for purchase from vendors, but this one is free for you to use.
3. To evaluate an actual acquisition project, I recommend using a professional modeling and migration package to be able to capture the complex velocities and amplitudes of your geologic environment.

An example of the starting user interface is shown in Figure A.1. It is divided into seven panels, each providing control of the modeling and migration features. The Model panel allows the dimensions of the model and the velocity information to be entered. The Well panel allows for the geometry of the borehole or surface fiber to be entered. The Reflector panel allows for the subsurface layers to be defined and controlled. The Simulation panel controls the forward modeling parameters of the source, wave types, gauge length, and trace sampling information. The Noise panel allows the forward-modeled data to be displayed and various levels of noise to be added to the data. Finally, the Migration panel allows the modeled data to be migrated.

A.1. Creating a model

After launching the ExploreDAS application, the first task is to set the data directory where all of your modeling results will be stored. In Figure A.2 the "File" menu item has been selected showing the "Set Data Directory" option.

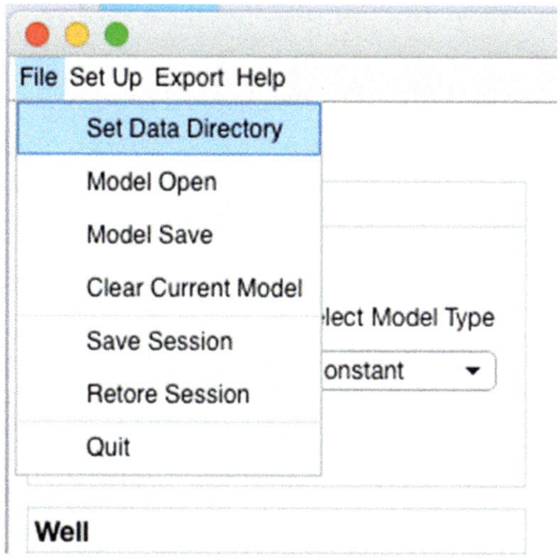

Figure A.2. File menu drop-down showing how to set the data directory.

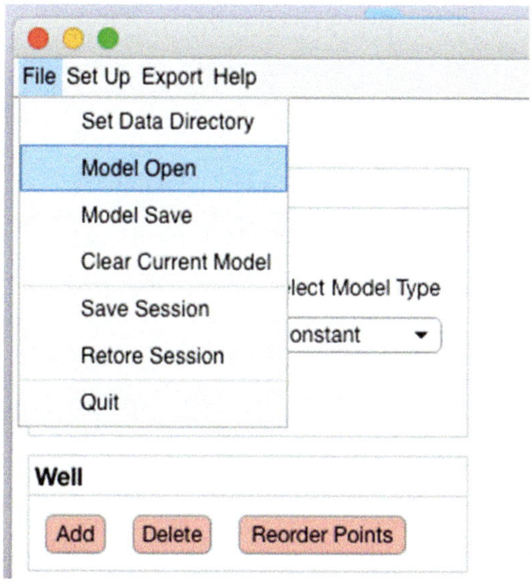

Figure A.3. File menu drop-down showing how to read in a model already saved on disk.

If you have already used ExploreDAS and saved a velocity model, then you can open the saved model from the File menu drop-down Model Open, as shown in Figure A.3.

If you want to start a new model from scratch, then enter the model dimensions in the Model panel, as shown in Figure A.4. The Xmin and Xmax editable fields set the lateral size of the model. The Zmin and Zmax editable fields set the depth range of the model. Both of these units are in meters. The drop-down selection box entitled "Select Model Type" allows you to select whether the velocity model will be constant or have a vertical gradient. When the "Constant" model type is selected, then the two fields labeled "P Velocity" and "S Velocity" allow you to set the constant velocity of the model. The units are in m/s.

However, if you select the "Gradient" option, then a different set of editable fields are displayed as shown in Figure A.5. In this display, the "Vp0" and "Vs0" fields show the initial velocity of the model for the P- and S-waves, respectively, of the depth = 0 m. The velocity gradient, kZ, is set in the corresponding editable field with units of 1/s.

A.2. Defining the well geometry

The Well panel allows you to create the trajectory for the fiber and geophone sensors. The geometry can be completely arbitrary. This means the "well" can be along the surface of the ground, or in the subsurface

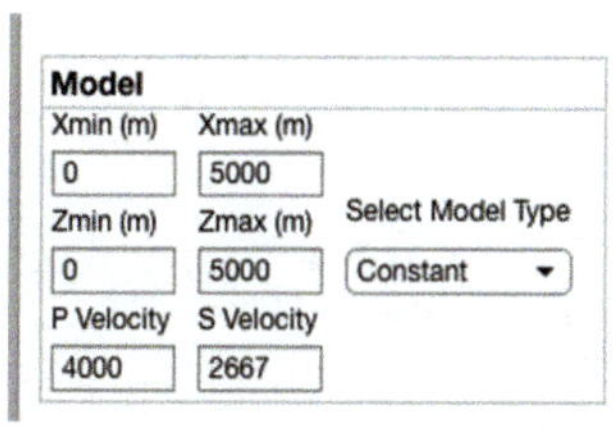

Figure A.4. Model panel allowing the size of the model and velocity information to be set.

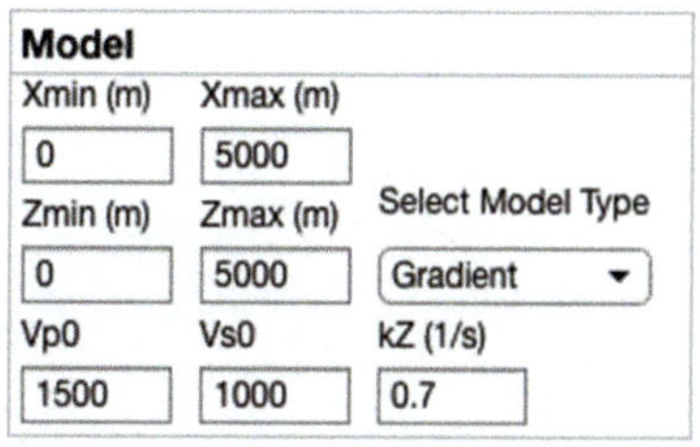

Figure A.5. Model panel allowing the velocity gradients to be set.

along a path that you digitize. You start by pushing the "Add" button in the Well panel, shown in Figure A.6.

After the "Add" button is pushed, you can start to digitize the well path by depressing the left mouse button at the locations in the model you would like to cover, as shown in Figure A.7. The label at the top of the model gives you directions on how to digitize the model. Right-click the mouse button (anywhere in the model) when you are finished digitizing the well trajectory. (The right mouse button click is one of those things users tend to forget which leaves the well trajectory incomplete.)

If you want to make a correction you can press the "Delete" button and wherever you click, it will search for the nearest point and delete it. On the other hand, if you want to insert more points, you can press the "Add" button again in the Well panel and start digitizing. When all the new points have been digitized, remember to right-click the mouse to stop adding points. The new points will be added after the last point that was digitized which will likely not be what you want to have happen, as shown in Figure A.8.

However, if you push the "Reorder Points" button, the digitized points will be reordered by sorting them starting with the first point originally digitized and then finding the nearest digitized point one after each other, as shown in Figure A.9.

The digitized points for the well trajectory can be of arbitrary spacing and the code will perform linear interpolation between the digitized points with a new spacing of 10 m when it is used for computing the seismic traces.

Figure A.6. The Well panel.

A.3. Creating the reflector geometry

Next the reflections are created. Note again that the model is either a simple constant velocity or simple linear gradient in depth. Therefore, there is no contrast in impedance – meaning that the reflections are all of the

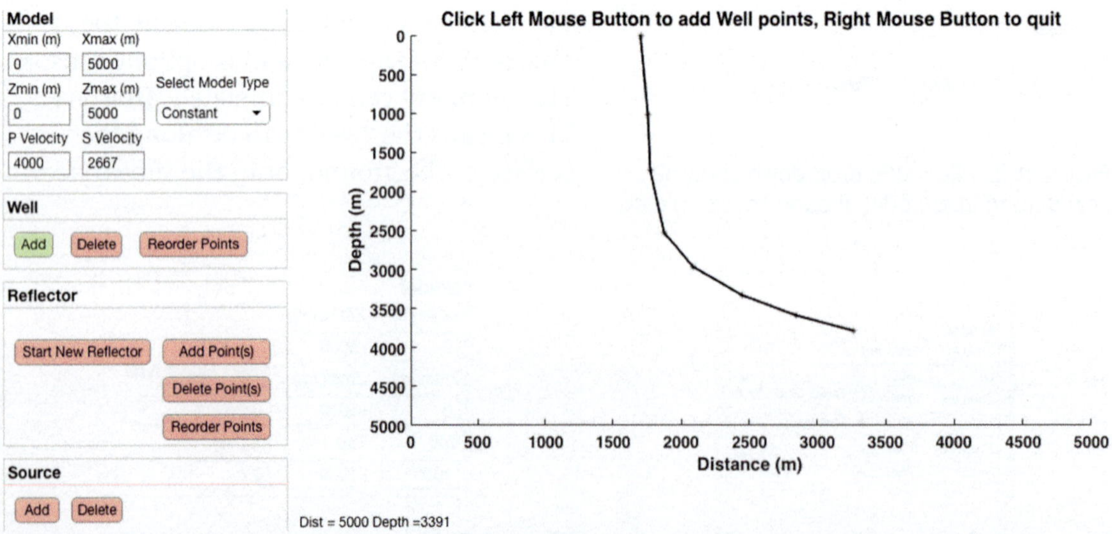

Figure A.7. The user interface while the well path is being digitized after pressing the "Add" button.

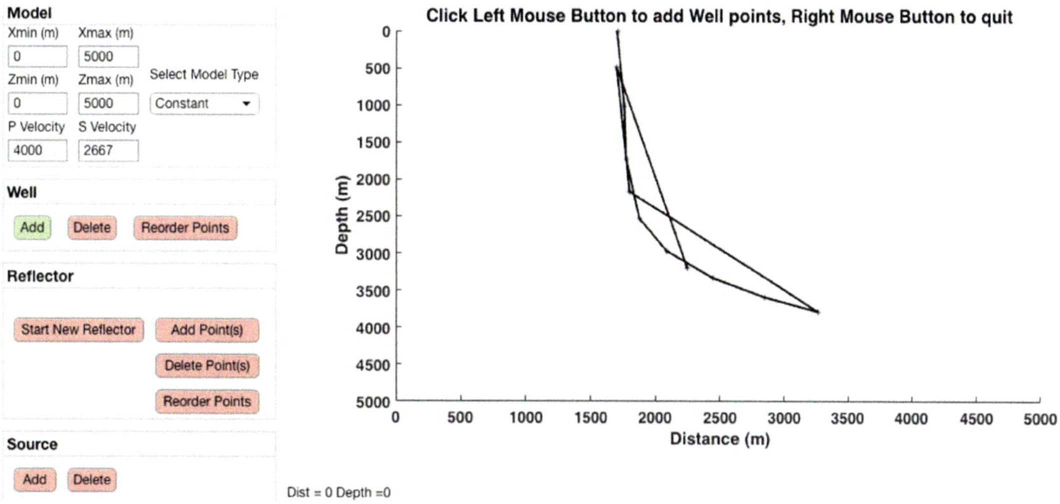

Figure A.8. The well trajectory after inserting 3 new points. This will require the "Reorder Points" button to be pressed to put them into a reasonable order.

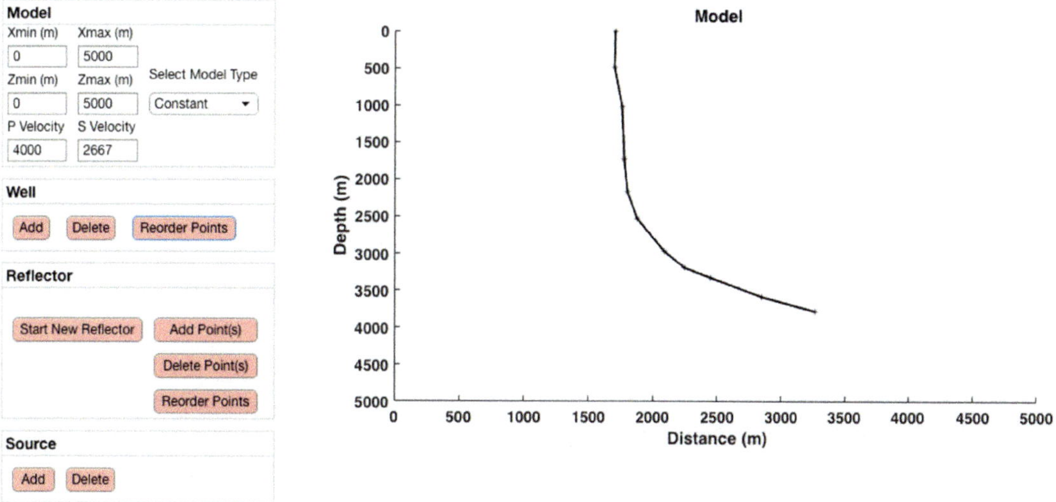

Figure A.9. The well trajectory after pressing the button "Reorder Points" in the Well panel.

same amplitude and only the kinematic (arrival times) are correct. To add the first layer, push the "Start New Reflector" button in the Reflector panel. When this button is pushed, it creates a new layer, for first time it is labeled "1" as shown by the drop down menu in the Reflector panel, as shown in Figure A.10.

To digitize points for this reflector, push the "Add Point(s)" button and digitize as shown in

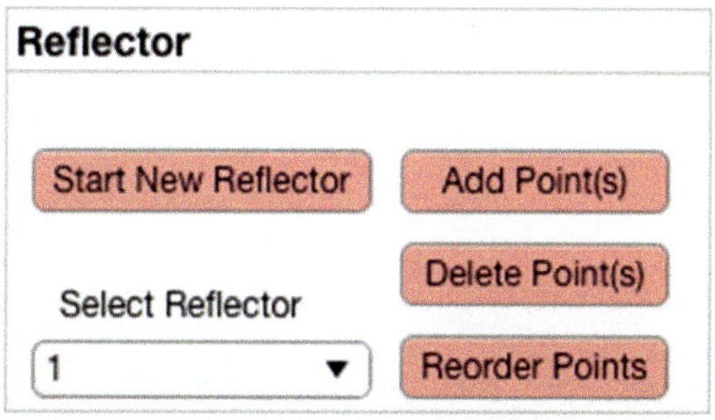

Figure A.10. Reflector panel after pushing "Start New Reflector" button and the first reflector is populated in the drop-down list.

Figure A.11, like you did for the well trajectory. Note that the current reflector being digitized will show as a blue line with the red asterisks as the digitized points. Remember again to finish digitizing by right-clicking the mouse button.

To add a new reflector, push the "Start New Reflector" button and repeat the operations just described.

You can modify the reflector points by first selecting the reflector number you want to modify in the drop-down list in the Reflector panel. Then, as you did with the well trajectory editing, push the "Delete Points(s)" button in the Reflector panel to remove unwanted points. New points can be added by pushing the "Add Points(s)" button and digitizing again. Pushing the "Reorder Points" button will again sort the reflection points into the best reasonable nearest neighbor order. Figure A.12 shows the result of adding 5 reflectors to the model.

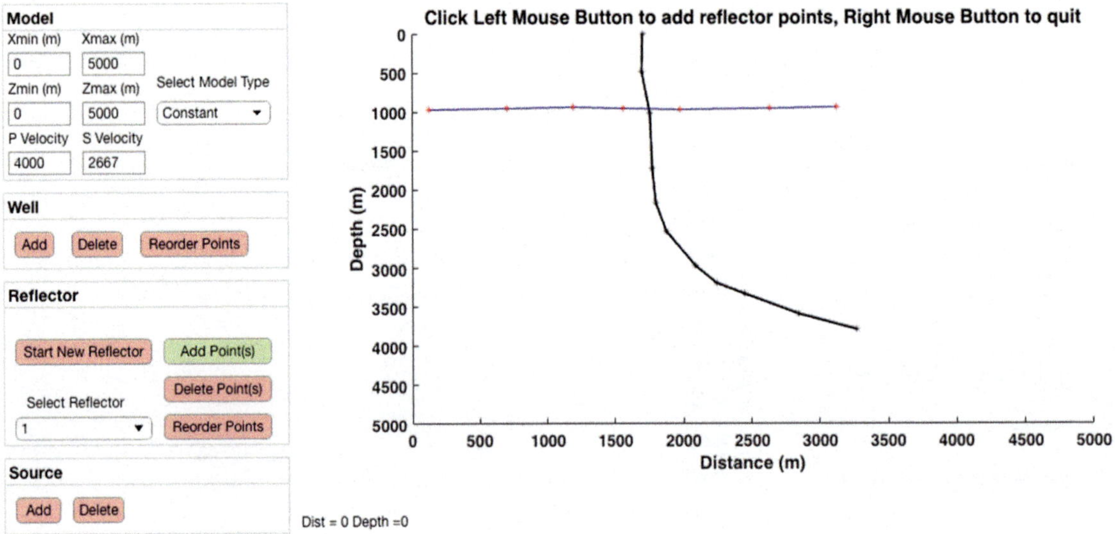

Figure A.11. The first reflector is being digitized.

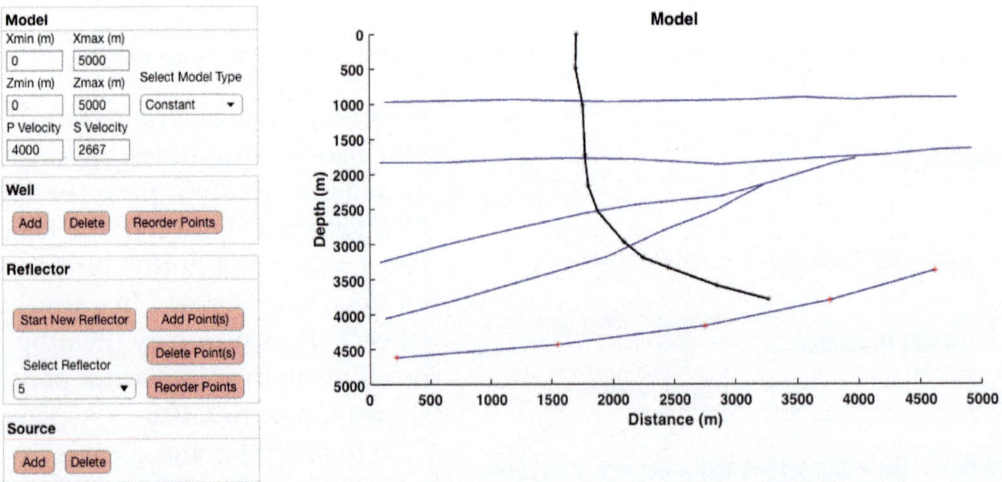

Figure A.12. The geometry of the well trajectory and reflectors after entering five reflectors.

The geometry of the reflectors you create is completely arbitrary. The layers do not have to extend across the entire model. They can cross each other and have any geometry. Of course, the more crazy you make the layers the less realistic the modeling results will be.

A.4. Creating the source geometry

Next the locations of the seismic sources are entered. You can have one or any number of sources. To create a source location, push the "Add" pushbutton in the Source panel. Then digitize wherever you want a source to be located. As before, remember to right-click the mouse button to stop entering in source locations. You can delete individual source locations by pushing the "Delete" pushbutton in the Source panel. Figure A.13 shows an example of three source locations digitized, as indicated by the red asterisks.

A.5. Digitizing from an image

I've described how you can create a model using "free hand" digitizing. However, if you have an image of a model or seismic section, you put that behind the model you are creating. To import a jpeg image, select the "Read in Image Behind Model" option from the "Setup" menu as shown in Figure A.14. It will ask you to select an image to

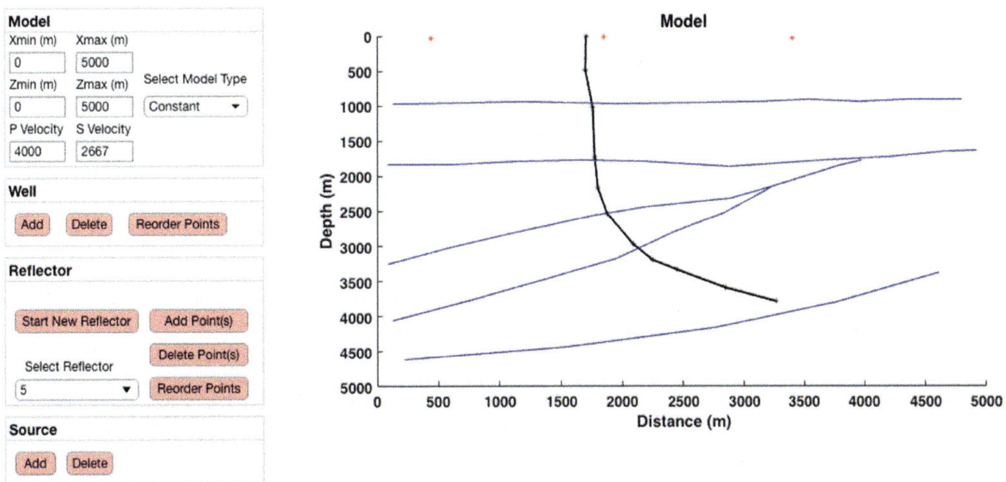

Figure A.13. Model after the well, reflectors, and three shot locations have been digitized.

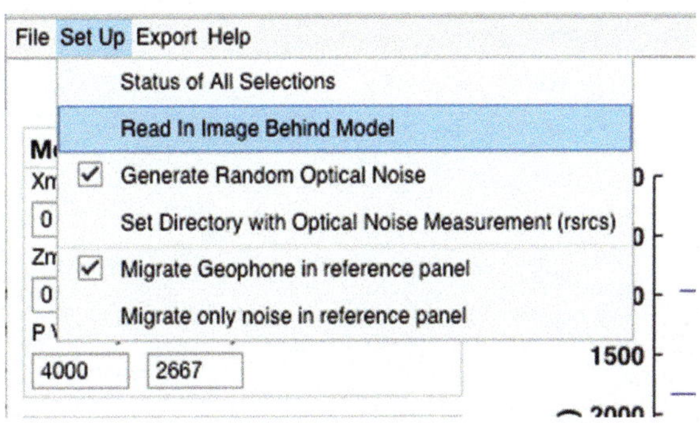

Figure A.14. Use the "Read In Image Behind Model" menu option to import an image to add in creating the model.

import. It then reads in that image, displays it in a new figure, as shown in Figure A.15, and asks you to digitize the upper left corner, then the lower right corner, corresponding to the dimensions of the model you have specified in the Model panel.

The model will appear in the user interface, as shown in Figure A.16, and you can start to create the Well, Reflector, and Source geometries. You can toggle off and on the image behind the geometry by pushing the "No Image" button in the Model panel.

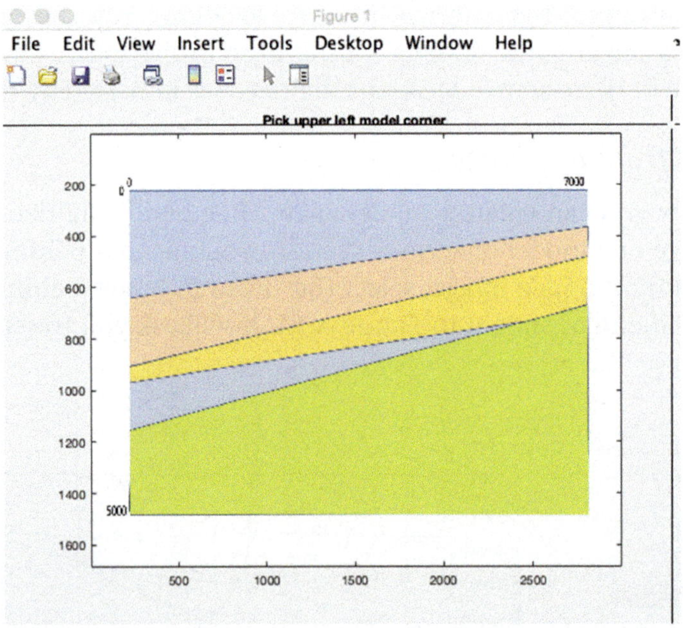

Figure A.15. After you select a jpeg file to use as a background, a new figure appears for you to digitize the upper left corner and lower right corner of the image corresponding to your model limits.

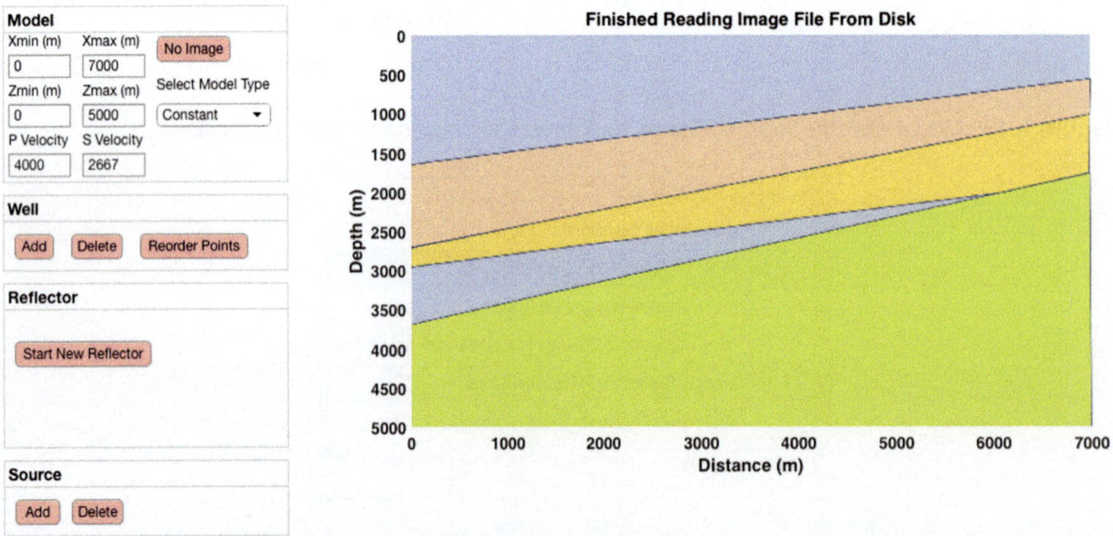

Figure A.16. User interface after reading in a background image.

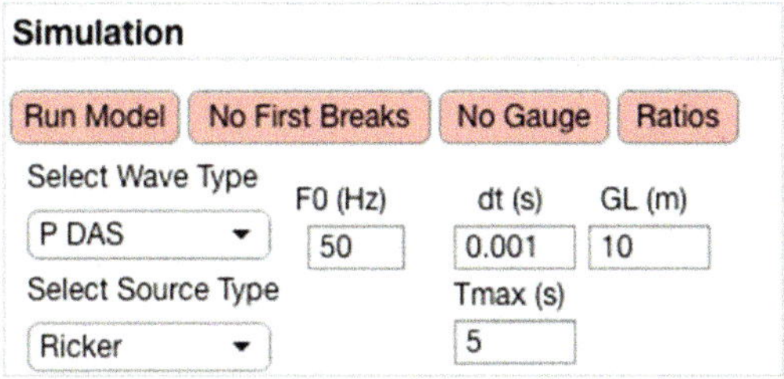

Figure A.17. The Simulation panel.

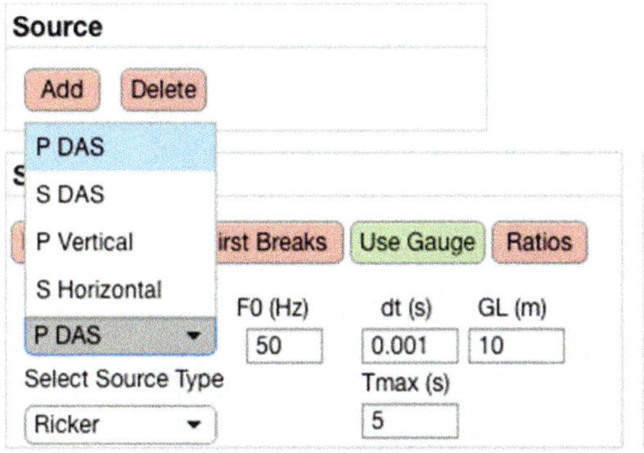

Figure A.18. The Select Wave Type drop-down list providing the choice of response to model.

A.6. Creating shot records

Next you will want to create shot records for the model you have built. The Simulation panel, in Figure A.17, provides choices for the forward modeling. The bottom left drop-down provides a choice of using a Ricker or Klauder wavelet. The Ricker wavelet has only one parameter and that is the center frequency of the source wavelet. For the Klauder wavelet, there are two parameters – the lower and upper frequency limits. The Klauder wavelet is broader band but has ringing side lobes. So, for "pretty" clean-looking data choose a Ricker source. But actually, the Klauder wavelet is more representative of a broad-band vibrator source.

You set the sampling interval of the seismic trace using the "dt" parameter specified in seconds. Most surface seismic data is recorded at 2 to 4 milliseconds (0.002 and 0.004 s, respectively). Most VSP and microseismic data is recorded at 1 millisecond (0.001 s) or smaller sampling interval. Thus providing a "dt" value of 0.001 s is a good compromise. The value of "Tmax" is the maximum time length of the seismic record in seconds. The value you want depends upon the velocities you choose, the depth of your model, the offset of your sources, and if you have surface or VSP type well geometry.

The GL parameter is the gauge length in meters. You will most likely want to test several different values to see what its effect will be on your data.

Across the top of the Simulation panel are four buttons. Pressing the "Run Model" button will start the forward modeling job. The "No First Breaks" button allows you to remove the first break events from the seismic record by not simulating the direct arrival. This is useful because for migration of field data you usually want to remove the first breaks. The label on the button indicates whether the first breaks are being removed or included. If it is red and reads "No First Breaks," the first breaks will not be computed. If you toggle it to green, it now reads "First Breaks," and the shot records will contain the first breaks.

The next button is labeled "No Gauge." In this state, it means that no gauge length will be applied to the shot records. It will have the appropriate angle of incidence response for fiber but not the gauge length. By toggling this button to green, the label will now read "Use Gauge" and the seismic data generated will have the gauge applied. The editable field titled "GL (m)" allows you to specify the desired gauge length.

The drop-down labeled "Select Wave Type" allows you to select the type of seismic data you want to create. The "P DAS" and "S DAS" options create the P-wave and S-wave responses for a DAS sensor, respectively. The "P Vertical" and "S Horizontal" provide the single component geophone response.

For this example, we will select the P DAS wave type, a Ricker wavelet, 70 Hz, use a 10 m gauge length, and the maximum time (Tmax) as 2.5 s. After you push the "Run Model" button in the Simulation panel, two windows will pop up. The first window will contain the shot record images. The left image will be of the reference P-wave response, similar to a geophone response, and the right image will be the P-wave DAS response, as shown in Figure A.19. The second window will display one trace from geophone and DAS records on top of each other for comparison, as shown in Figure A.20.

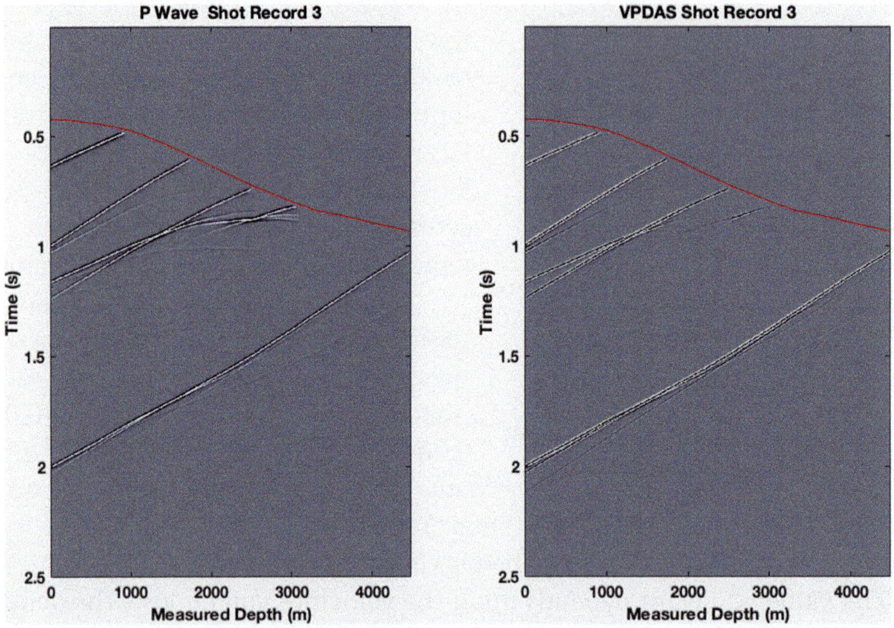

Figure A.19. A P-wave seismic shot record modeled for (left) geophone, and (right) DAS.

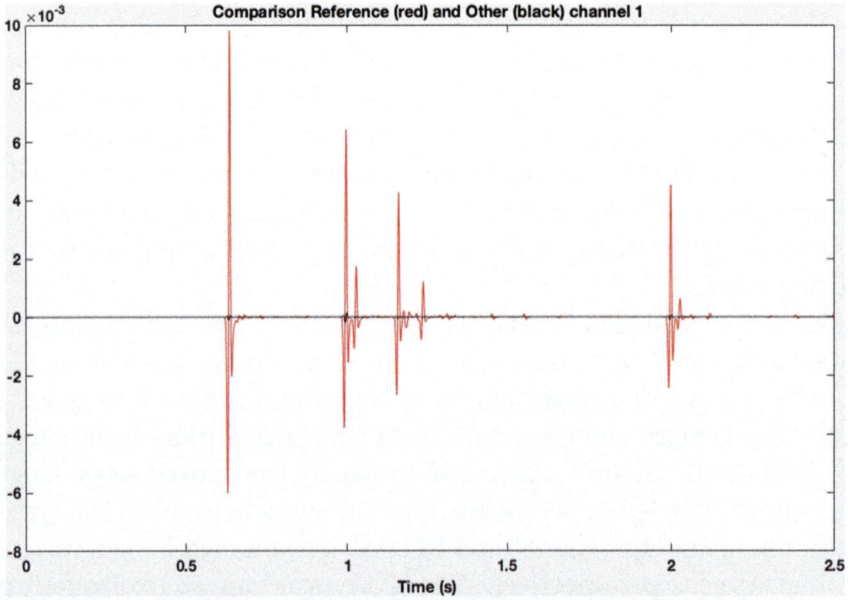

Figure A.20. Single trace comparison of the geophone (red) and DAS (black) simulations.

A.7. Viewing the shot records and adding noise

Once you have created the shot records by pushing the "Run Model" button in the Simulation panel, the Noise Parameter panel will fully populate with the options as shown in Figure A.21.

From the Source drop-down list, you can select the source number and the channel number from the Channel drop-down list. Once they have been selected, push the "Replot" button and the same two figures, as in Figures A.19 and A.20, will appear again. You can increase or decrease the plot gain by pushing the "Gain Up" or "Gain Dn" pushbuttons.

You can add simulated optical noise to the DAS data by toggling the "No Optical Noise" button to "+ Optical Noise." Below that button is an editable field with a title "Optical SNR" where you can specify the amount of optical noise to add in dB. In a similar fashion, you can add common mode noise (CMN) to the DAS data by toggling the "No CMN Noise" to "+ CMN Noise." Below that button is an editable field titled "CMN SNR" where you specify the amount of common mode noise to add in dB. Figure A.21 shows the shot records after adding both

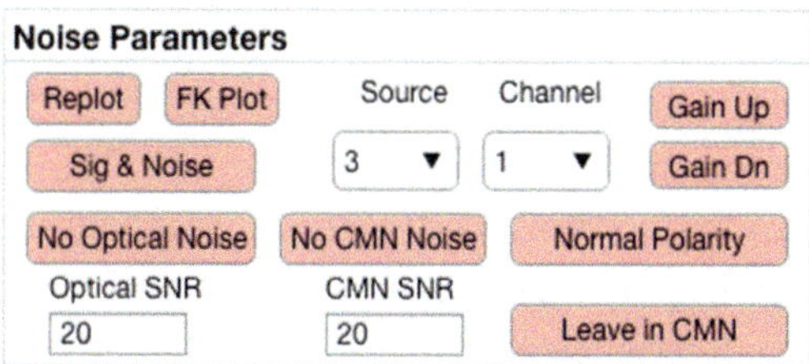

Figure A.21. The fully populated Noise Parameters panel.

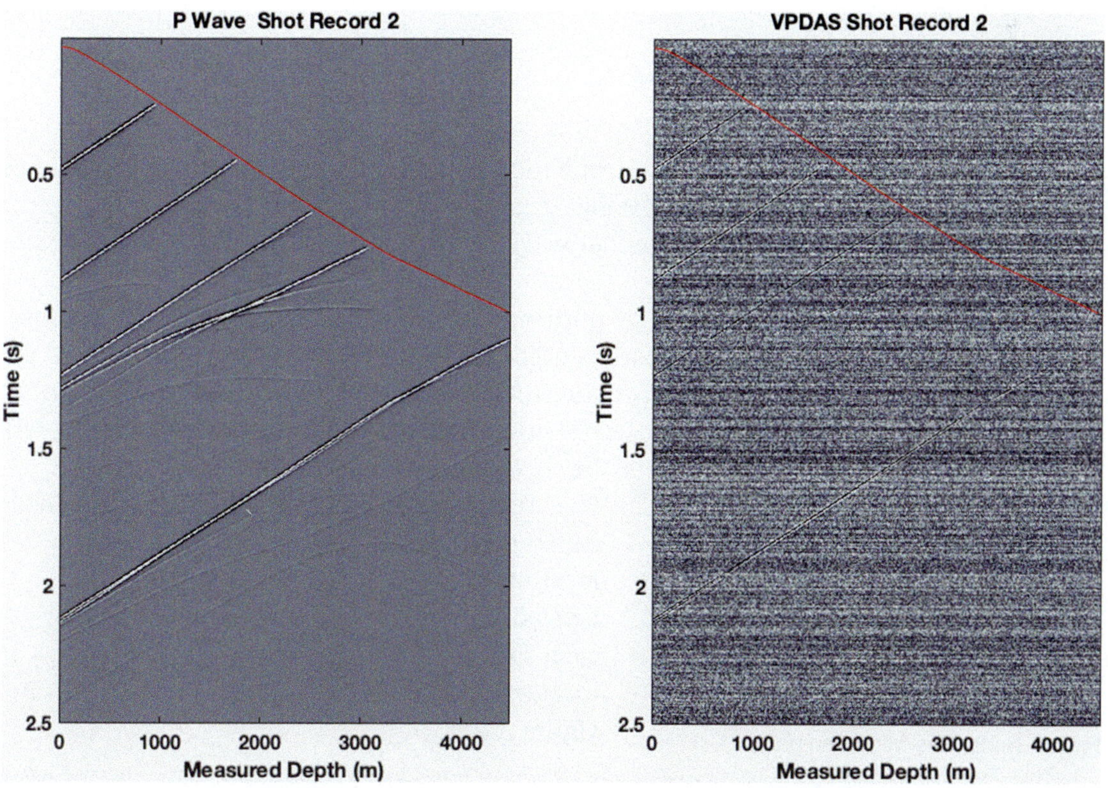

Figure A.22. Result of adding optical and common mode noise to the DAS record (right).

optical noise and common mode noise. You can remove the common mode noise via processing, like you would for field data, by toggling the "Leave in CMN" to "Remove CMN."

The polarity of the DAS record can be reversed by toggling the "Normal Polarity" push button to "Reversed Polarity." Another option available is to only plot the noise in the DAS data and not include the modeled DAS record. This is performed by toggling the "Sig & Noise" push button to "Noise Only." The final option in the Noise Parameters panel is the "FK Plot" push button. If you depress this button it will display the FK transform of the geophone and DAS records in new figure windows.

A.8. Migrating the data

The final step is to see the quality of the migrated image created from your modeled shot records. Figure A.23 shows the Migration panel. The first step in performing a migration of the data is to define the output image location and spacing. The editable fields labeled "Xmin (m)," "Xmax (m)," and "dX (m)" allow you to define the starting and ending lateral positions, as well as the trace spacing. The editable fields labeled "Zmin (m)," "Zmax (m)," and "dZ (m)" allow you to define the starting and ending depth positions, as well as the trace depth sampling interval. Depending on the frequency content of the source you have chosen, you need to select depth sampling interval (*dZ*) to be small enough to capture an unaliased signal. A general rule is to use Equation A.1 to pick *dZ*.

$$dZ < \frac{2.5C_{min}}{f_{max}} \tag{A.1}$$

where C_{min} is the minimum velocity in the model, and f_{max} is the maximum frequency of the source wavelet. For a Ricker wavelet, f_{max} can be set at twice the center frequency of the wavelet. For the Klauder wavelet, the specified maximum frequency can be used.

You start the migration process by pushing the "Migrate" pushbutton. All of shot records will be migrated. As each record is used, it will be shown in a new figure and the intermediate results will be displayed. When the migration has finished, the final images will be shown as in Figure A.24.

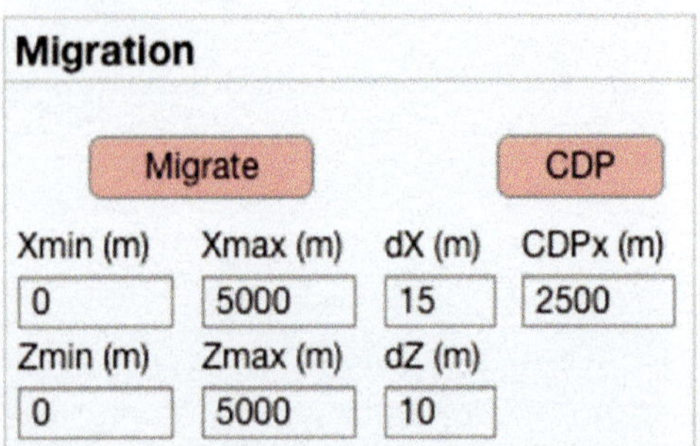

Figure A.23. The Migration panel.

You can also create a migrated common depth point (CDP) gather. You specify the location of the CDP using the editable field labeled "CDPx (m)." The migrated results for each shot will be created for this CDP location and displayed as a gather, as shown in Figure A.25.

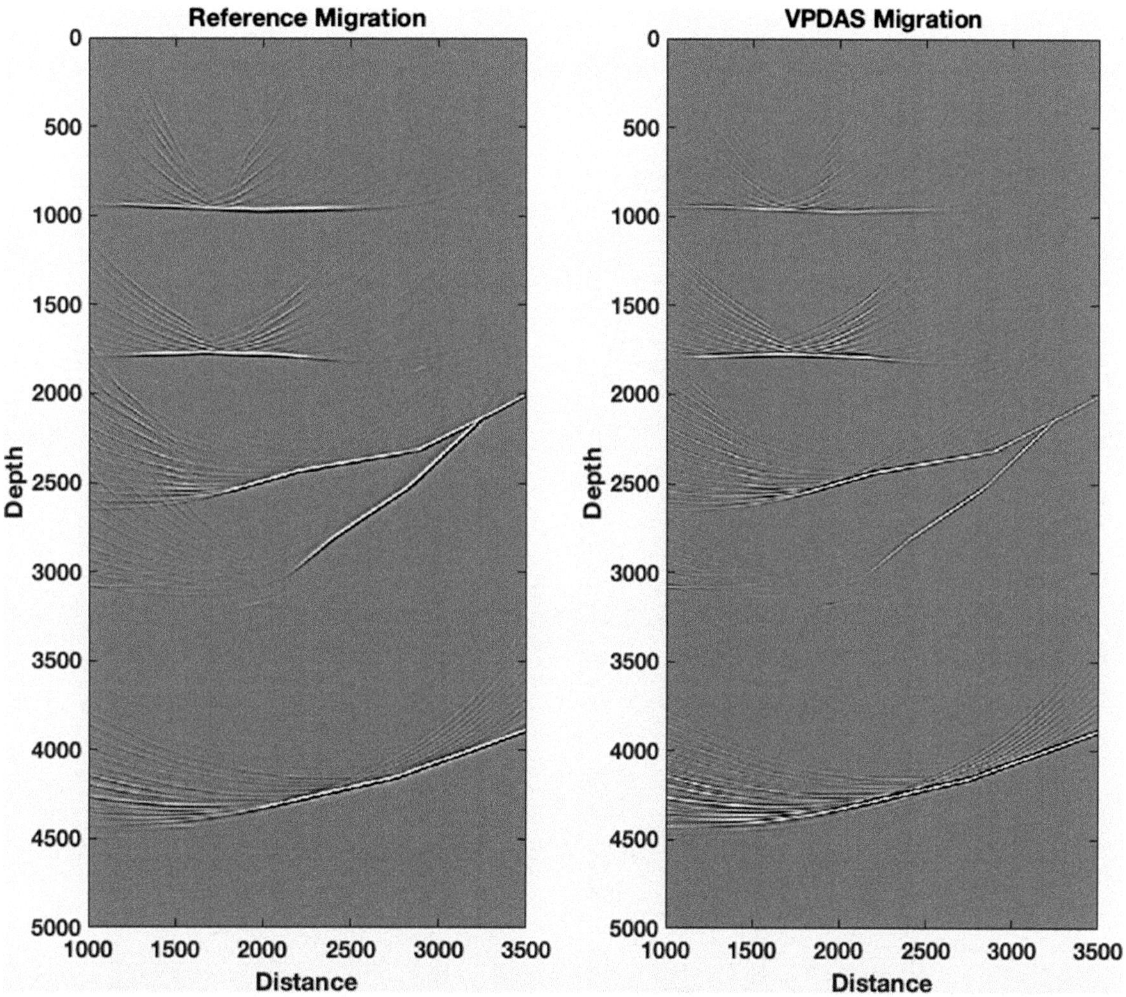

Figure A.24. The migrated results of 12 shots for (left) the geophone data, and (right) DAS data.

A.9. Saving the results

There are several ways to save your work in ExploreDAS. The first option is to save your model. This is done by selecting the "Model Save" option in the "File" menu drop-down, as shown in Figure A.26. You then select the directory for storing the model in JSON format using the file name you choose. If you want to start again later with this model, simply start up ExploreDAS and select the "Model Open" option in the "File" menu. Then select the JSON formatted file you created with the model.

To save your current working session environment, select the "Save Session" option in the "File" menu. You can restore your session at a later date by selecting the "Restore Session" option in the "File" menu and then opening the file you saved in the previous step.

An easy way to summarize the work you've done and view the settings you have made is to select the "Status of All Selections" option from the "Setup" menu option, as

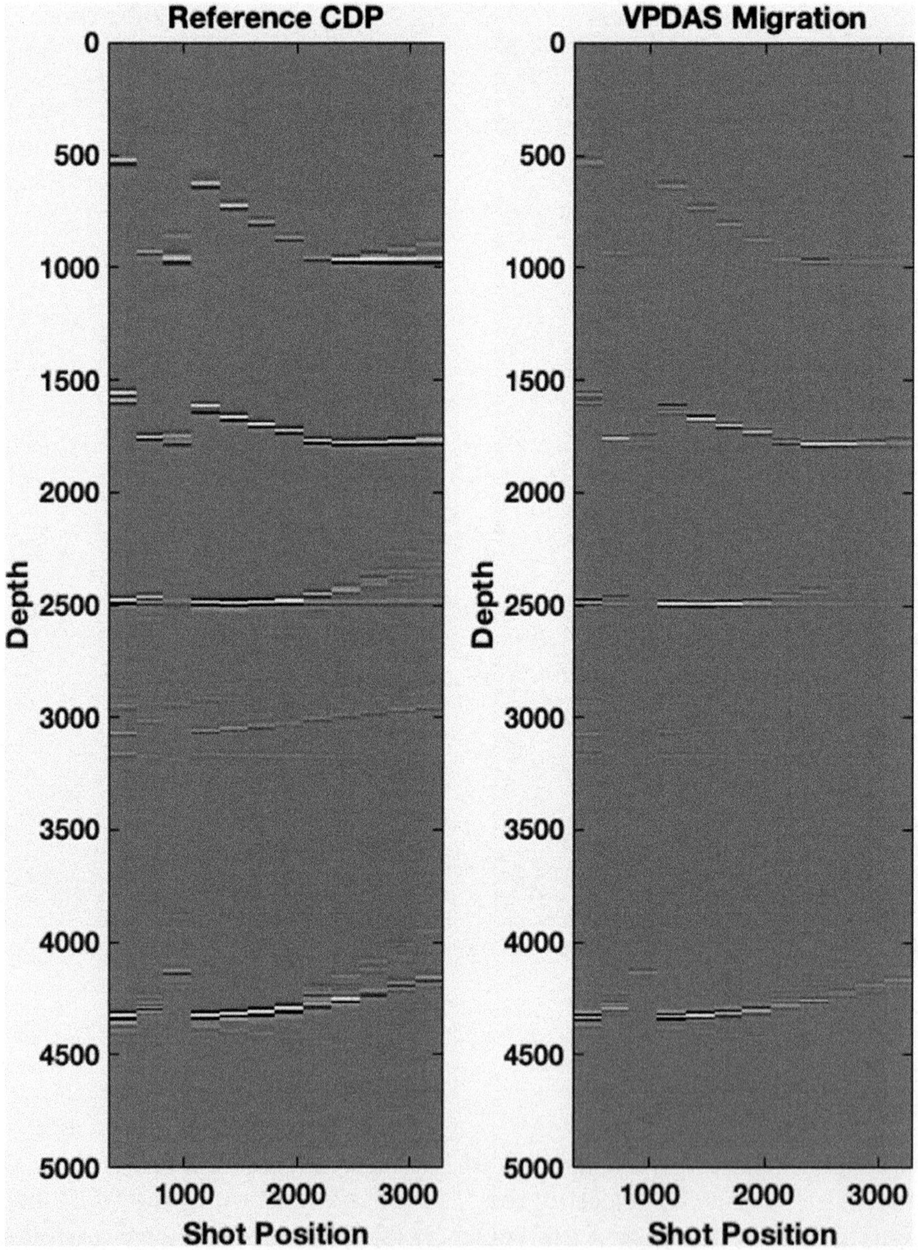

Figure A.25. Migrated CDP gather for (left) geophone, and (right) DAS data.

shown in Figure A.27. After you make this selection, a new window will appear that summarizes all of your work, as shown in Figure A.28.

Another option is to export the modeled data either to MATLAB or SEGY formatted files. (For writing SEGY formatted files, I've used the open-source release of SegyMAT 1.5.1 available from https://segymat.readthedocs.io/en/latest/ written by Thomas Mejer Hansen. I am very grateful for Thomas making his software freely available as it has proven extremely useful for many of my projects.) To export the shot records, migrated images, and/or CDP gathers select one of the options on the "Export" menu drop-down list, shown in Figure A.29.

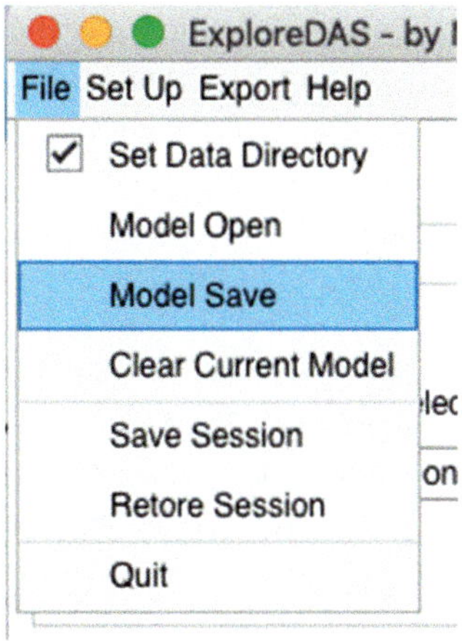

Figure A.26. The "Model Save" option will create a file with your model.

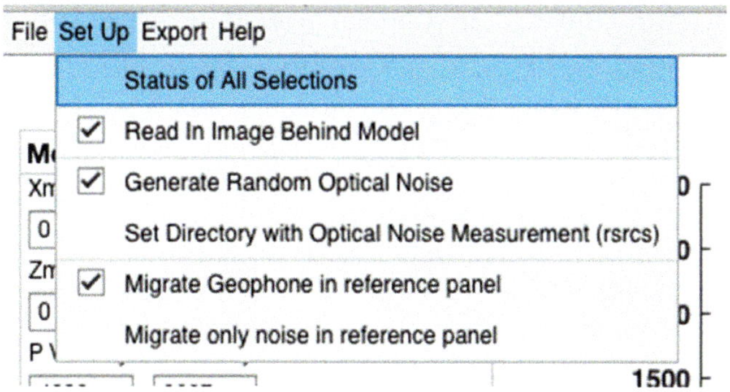

Figure A.27. Status of All Selections menu option.

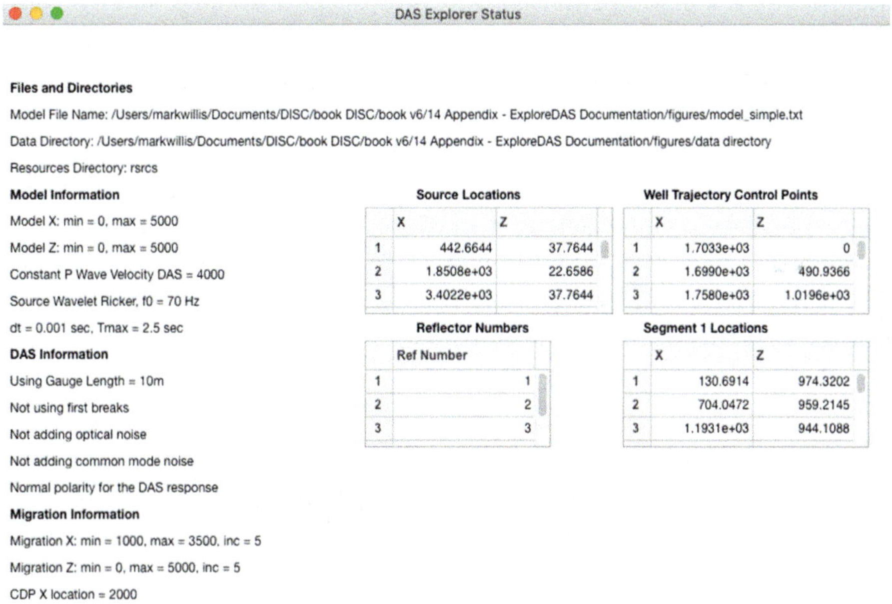

Figure A.28. The window showing all of your selections for this session.

Figure A.29. Use the "Export" menu to write out the data sets in either SEG or MATLAB formats.

A.10. Theory behind the modeling

If you have made it this far in the documentation, I commend you. In my experience, most users are way too happy to use software without reading the directions and are loathe to trudge through the documentation. Hopefully you have found the user interface of ExploreDAS to be intuitive enough that you can build models easily and explore how DAS data sets are useful. However, if you find yourself reading this section, I'm guessing that you have decided that you are questioning the validity or wanting to know the limitations of this modeling and migration software. So, I'll explain how the software does what it does.

A.10.1 Model well trajectory

You may have noticed that the well trajectory is completely arbitrary. The well can make loop to loops and sharp turns as shown in Figure A.30. The points you digitize are used on the fly, to interpolate to a more finely spaced depth interval. However, there is no smoothing or curve fitting through the points; thus, there will be some artifacts in the data created with crazy geometries like the one shown in Figure A.30.

A.10.2 Model reflector geometry

In a similar fashion to the well trajectory, the reflector geometry is not enforced to be regular or conformal. The reflectors do not have to go from edge to edge of the model. The reflectors do not have to define enclosed regions. Because the velocity model is simple, either constant or v(z), the reflector positions just create locations that scatter (and reflect) the incident seismic energy. Figure A.31 shows a model with three simple planar layers and a crazy loop-to-loop layer. Before each reflector is used in the modeling, it is more finely interpolated so that there are many scattering points along it. Because the modeling is performed by scattering off of each of the reflector points, this means there will be artifacts generated from the first and last points of each reflector. Sharp corners will also cause scattering. The software does not attempt to correct or modify any of the features you put into the model. So your synthetic shot records will be just as valid as the model you create.

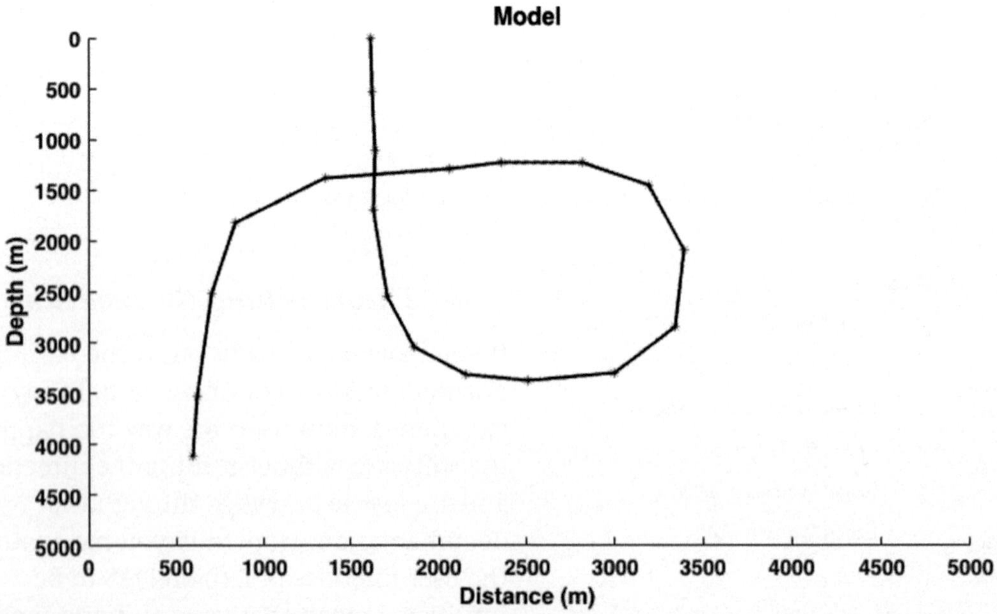

Figure A.30. Well trajectory that is absolutely unreasonable and crazy!

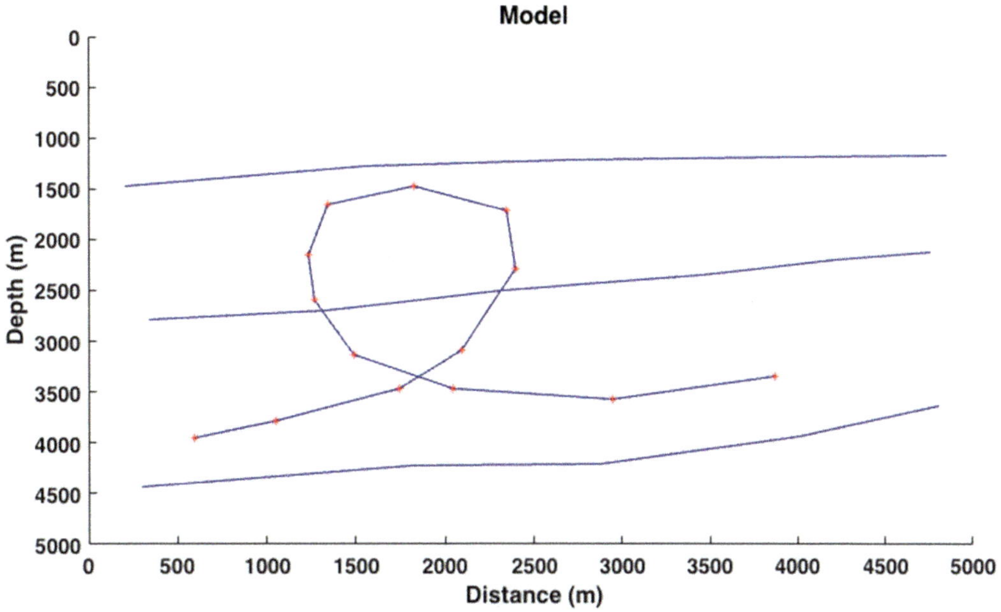

Figure A.31. A four-reflector model with one crazy reflector.

A.10.3 Generation of shot records

The geophone shot records are created by simple scattering of each point of the reflector (Willis et al., 2015). For a trace associated with a given pair of shot and receiver locations, the traveltime is computed from the shot to a point on a reflector and then onward to the receiver. This traveltime is used to delay the source wavelet and is summed into the trace. The contribution from each point along each reflector is added into the trace. The current code assigns the same unity reflection coefficient for each reflector.

The vertical and horizontal geophone components, if selected for output from the drop-down list, are computed using the appropriate $\cos(\theta)$ (P-wave) or $\sin(\theta)$ (S-wave) weighting for the angle of incidence at the well.

To compute the DAS traces, the geophone response is computed for a pair of locations about the receiver (or channel) location space apart by the gauge length. Then the difference of the two traces are taken. Then the appropriate $\cos^2(\theta)$ (P-wave) or $\sin(2\theta)$ angle of incidence weighting is performed for the angle of incidence at the well.

A.11 Theory of migration

To create the migrated images, a simple Kirchhoff depth migration is performed (Hill and Ruger, 2019). I'll give an extremely quick overview of Kirchhoff migration. (Those new to migration will be totally confused, and those who are experts will be upset that I explained it poorly. Sorry to both groups! Hopefully somebody will be happy with my rapid explanation.) Each trace in the data set has a source position and a receiver location, as shown in Figure A.32. Each trace is migrated into the image separately. For each point in the migrated image, the traveltime from the source to that image point (magenta line) is added to the traveltime from that image point to the receiver (blue line). The sum of the two traveltimes gives the time in the trace (red arrow) from which

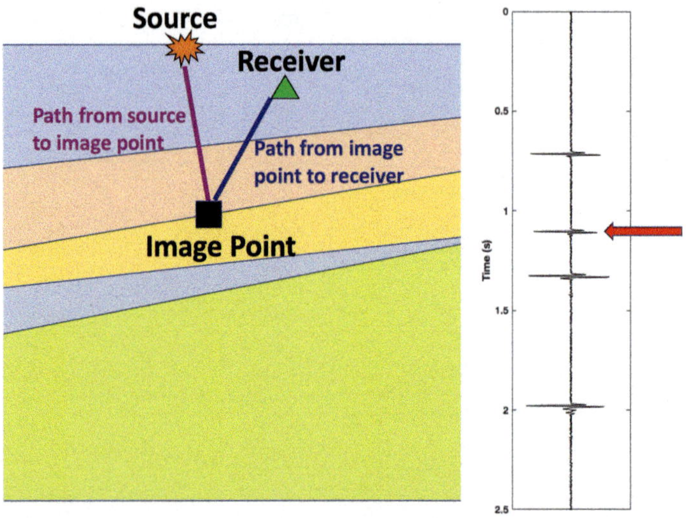

Figure A.32. (Left) The subsurface showing the geometry for a source (red), receiver (green), and image point (black). (Right) The seismic trace from this source and receiver location. The red arrow shows the computed time from the source, to an image point, and to the receiver.

is extracted the amplitude of the trace at that time. This amplitude value is added to the point in the evolving migrated image. Thus, each trace contributes to all samples in the migrated image. Constructive interference will reveal the reflectors in the subsurface. This is simplicity at its finest. I have not implemented any fancy weighting or filtering.

A.12 Summary

This open-source software package is being freely provided to you with the desire that it proves useful and instructive. There are many improvements that could be made, for example: adding ray tracing for more complex models, the addition of anisotropy, the addition of the response of helically wound fibers, etc. If you are so inspired, I hope you will take up the initiative and share your results. However, even in its current form I believe ExploreDAS will be sufficient for you to learn about the power that DAS data sets provide to learn more about the earth's subsurface.

References

Ajo-Franklin, J., S. Dou, N. Lindsey, I. Monga, and C. Tracy, 2019, Distributed acoustic sensing using dark fiber for near-surface characterization and broadband seismic event detection: Nature Scientific Reports, doi: 10.1038/s41598-018-36675-8.

Ajo-Franklin, J., S. Dou, N. Lindsey, T. Daley, B. Friefeld, E. Martin, M. Robertson, C. Ulrich, T. Wood, I. Eckblaw, and A. Wagner, 2017, Timelapse surface wave monitoring of permafrost thaw using distributed acoustic sensing and a permanent automated seismic source: 87th Annual International Meeting, SEG, Expanded Abstracts, 5223–5227, doi: 10.1190/segam2017-17774027.1.

Alfataierge, E., A. Aldawood, A. Bakulin, and R. Stewart, 2020, Influence of gauge length on DAS VSP data at the Houston Research Center test well: 90th Annual International Meeting, SEG, Expanded Abstracts, 505–509, doi: 10.1190/segam2020-3419066.1.

Ali, S., F. Miranda, S. Emmanuel, and R. Castaneda-Aguilar, 2021, Processing of a large offshore 3DVSP DAS survey in a producing well: First International Meeting for Applied Geoscience and Energy, SEG/AAPG, Expanded Abstracts, doi: 10.1190/segam2021-3594397.1.

Amira, Z., M. Bouyahi, and T. Ezzedine, 2015, Measurement of temperature through Raman scattering: International Conference on Advanced Wireless Information and Communication Technologies (AWICT 2015), 350–357, doi: 10.1016/j.procs.2015.12.003.

Ander, M., W. Kerr, C. Aiken, C. Glover, and M. Zumberge, 1990, An absolute wireline calibration to support a test of Newton's inverse square law: Geophysics, 55, no. 7, 920–923, doi: 10.1190/1.1442907.

Anstey, N., 1986, Whatever happened to ground roll?: The Leading Edge, 5, no. 3, 40–45, doi: 10.1190/1.1439239.

Baihly, J. D., C. Wilson, A. Menkhaus, E. Ejofodomi, K. England, and P. Dickerson, 2015, Fiber-optics revelations from a multistage openhole lateral fracturing treatment: URTeC 2172404, doi: 10.15530/URTEC-2015-2172404.

Baird, A., 2020, Modelling the response of helically wound DAS cables to microseismic arrivals: First EAGE Workshop on Fibre Optic Sensing.

Bakku, S., 2015, Fracture characterization from seismic measurements in a borehole: PhD thesis, MIT.

Bakulin, A., I. Silvestrov, and R. Pevzner, 2020, Surface seismics with DAS: An emerging alternative to modern point-sensor acquisition: The Leading Edge, 39, no. 11, 808–818, doi: 10.1190/tle39110808.1.

Bakulin, A., P. Golikov, R. Smith, K. Erickson, I. Silvestrov, and M. Al-Ali, 2017, Smart DAS upholes for simultaneous land and near-surface characterization and subsurface imaging: The Leading Edge, 36, no. 12, 1001–1008, doi: 10.1190/tle36121001.1.

Blount, C., K. Friehauf, B. Smith, D. Smith, M. Jaakelainen, and C. Baldwin, 2015, Improving fiber-optic system reliability in permanent installations: Society of Petroleum Engineers, SPE-174928-MS, doi: 10.2118/174928-MS.

Bolognini, G., J. Park, M. Soto, N. Park, and F. Di Pasquale, 2007, Analysis of distributed temperature sensing based on Raman scattering using OTDR codeing and discrete Raman amplification: Measurement Science & Technology, 18, no. 10, 3211–3218, doi: 10.1088/0957-0233/18/10/S24.

Bona, A., S. Ziramov, R. Pevzner, K. Tertyshnikov, and M. Urosevic, 2021, Distributed acoustic sensing for surface seismic: Case study from mineral exploration: 82nd Conference and Exhibition, EAGE, Extended Abstracts, doi: 10.3997/2214-4609.202011593.

Bona, A., T. Dean, J. Correa, R. Pevzner, K. Tertyshnikov, and L. Van Zaanen, 2017, Amplitude and phase response of DAS receivers: 79th Conference and Exhibition, EAGE, Extended Abstracts, doi: 10.3997/2214-4609.201701200.

Borodin, I., and A. Segal, 2020, Real-time hydraulic fracture monitoring and wellbore characterization with distributed acoustic sensing of pumping noise: The Leading Edge, 39, no. 11, 785–792, doi: 10.1190/tle39110785.1.

Bracewell, R., 1965, The Fourier Transform and its Applications: McGraw-Hill Electrical and Electronic Engineering Series: McGraw-Hill Book Company.

Bragato, P., P. Comelli, A. Saraò, D. Zuliani, L. Moratto, V. Poggi, G. Rossi, C. Scaini, M. Sugan, C. Barnaba, P. Bernardi, M. Bertoni, G. Bressan, A. Compagno, E. Del Negro, P. Di Bartolomeo, P. Fabris, M. Garbin, M. Grossi, A. Magrin, E. Magrin, D. Pesaresi, B. Petrovic, M. P. P. Linares, M. Romanelli, A. Snidarcig, L. Tunini, S. Urban, E. Venturini, and S. Parolai, 2021, The OGS-Northeastern Italy Seismic, and Deformation Network: Current status and outlook: Seismological Research Letters, 92, no. 3, 1704–1716, doi: 10.1785/0220200372.

Cai, Z., G. Yu, Q. Zhang, Y. Zhao, Y. Chen, Y. Jin, and H. Zhao, 2016, Comparative research between DAS-VSP and conventional VSP data: SEG Rock Physics & Borehole Geophysics Workshop, doi: 10.1190/RP2016-022.

Carpenter, C., 2016, Distributed acoustic sensing for downhole production and injection profiling: Journal of Petroleum Technology, Feb. 29, doi: 10.2118/0316-0078-JPT.

Chambers, K., 2020, Using DAS to investigate traffic patterns at Brady Hot Springs, Nevada, USA: The Leading Edge, 39, no. 11, 819–827, doi: 10.1190/tle39110819.1.

Chang, J., S. de Ridder, and B. Biondi, 2016, High-frequency Rayleigh-wave tomography using traffic noise from Long Beach, California: Geophysics, 81, no. 2, B43–B53, doi: 10.1190/geo2015-0415.1.

Chen, Y.-Z., G.-M. Hu, J.-J. Wu, G. Yu, Y.-P. Li, J.-H. Huang, S.-Z. Wang, and F. Li, 2021, Denoising analysis and processing methods of distributed acoustic sensing (DAS) vertical seismic profiling (VSP) data, *in* Y. Li, M. Karrenbach, and J. Ajo-Franklin, and the American Geophysical Union, eds., Distributed Acoustic Sensing in Geophysics: Methods and Applications, 1st ed. Geophysical Monograph 268: John Wiley & Sons Inc., doi: 10.1002/9781119521808.ch7.

Chen, J., J. Wang, J. Zhang, W. Chen, and W. Xu, 2020, Adaptive distributed acoustic sensing coupling noise removal method: 2nd SEG Borehole Geophysics Workshop, doi: 10.1190/bhgp2020-24.1.

Chen, J., J. Ning, W. Chen, X. Wang, W. Wang, and G. Zhang, 2019, Distributed acoustic sensing coupling noise removal based on sparse optimization: Interpretation, 7, no. 2, T373–T382, doi: 10.1190/INT-2018-0080.1.

Chen, J., W. Chen, X. Wang, Y. Zhou, Z. Shi, and G. Zhang, 2018, DAS coupling noise suppression using wavelet and DCT dictionary based on sparse optimization: 88th Annual International Meeting, SEG, Expanded Abstracts, 4938–4942, doi: 10.1190/segam2018-2996038.1.

Constantinou, A., A. Farahani, T. Cuny, and A. Hartog, 2016, Improving DAS acquisition by real time monitoring of wireline cable coupling: 86th Annual International Meeting, SEG, Expanded Abstracts, 5603–5607, doi: 10.1190/segam2016-13950092.1

Corning, 2021, Corning® SMF-28e+® Optical Fiber Product Information, doi: 10.2118/0316-0078-JPT, accessed 13 June 2022.

Correa, J., A. Egorov, K. Tertyshnikov, A. Bona, R. Pevzner, T. Dean, B. Freifield, and S. Marshall, 2017, Analysis of signal to noise and directivity characteristics of DAS VSP at near and far offsets – a CO2CRC Otway Project data example: The Leading Edge, 36, 12, 994a1–994a7, doi: 10.1190/tle36120994a1.1.

Coulter, K., and K. Goodman, 1991, Stretch corrected wireline depth measuring error and log quality indicator method and apparatus: U.S. Patent 5,062,048.

Cuny, T., P. Bettinelli, J. Le Calvez, R. Parker, R. Guerra, and M. Williams, 2020, Variable gauge length processing: DAS VSP examples: First EAGE Workshop on Fibre Optic Sensing, doi: 10.3997/2214-4609.202030008.

Daley, T. M., D. E. Miller, K. Dodds, P. Cook, and B. M. Freifield, 2016, Field testing of modular borehole monitoring with simultaneous distributed acoustic sensing and geophone vertical seismic profiles at Citronelle, Alabama: Geophysical Prospecting, 64, no. 5, 1318–1334, doi: 10.1111/1365-2478.12324.

Dean, T., K. Tertyshnikov, R. Pevzner, G. Calbris, C. Jestin, and V. Lanticq, 2019, Experimental measurement of the effects of acquisition parameters on DAS data quality: 81st Conference and Exhibition, EAGE, Extended Abstracts, doi: 10.3997/2214-4609.201901245.

Dean, T., T. Cuny, and W. Hartog, 2018, Gauge length optimization in distributed vibration sensing. U.S. Patent Application 2018/0003550 A1.

Dean, T., T. Cuny, and A. Hartog, 2017, The effect of gauge length on axially incident P-waves measured using fibre optic distributed vibration sensing: Geophysical Prospecting, 65, no. 1, 184–193, doi: 10.3997/2214-4609.201600684.

Dean, T., T. Brice, A. Hartog, E. Kragh, D. Molteni, and K. O'Connell, 2016a, Distributed vibration sensing for seismic acquisition: The Leading Edge, 35, no. 7, 600–604, doi: 10.1190/tle35070600.1.

Dean, T., A. Hartog, T. Cuny, and F. Englich, 2016b, The effects of pulse width on fibre-optic distributed vibration sensing data: 78th Conference and Exhibition, EAGE, doi: 10.3997/2214-4609.201600684.

Dean, T., B. Papp, and A. Hartog, 2015a, Wavenumber response of data recording using distributed fibre-optic systems: Third EAGE Workshop on Borehole Geophysics, doi: 10.3997/2214-4609.201412215.

Dean, T., T. Cuny, and A. Hartog, 2015b, Determination of the optimum gauge length for borehole seismic surveys acquired using distributed vibration sensing: 77th Conference and Exhibition, EAGE, Extended Abstracts, doi: 10.3997/2214-4609.201412740.

Duan, P., L. Le, C. Zhonglin, W. Guangyin, C. Yuanzhong, W. Junjun, H Guanming, and J. Gegand, 2021, Application of beam migration to DAS-VSP seismic data: 82nd Conference and Exhibition, EAGE, Extended Abstracts, doi: 10.3997/2214-4609.202112595.

Eaid, M., S. Keating, and K. Innanen, 2020a, The role of fiber geometry and gauge length in multiparameter elastic FWI of coiled DAS fiber data: 90th Annual International Meeting, SEG, Expanded Abstracts, 905–909, doi: 10.1190/segam2020-3427798.1.

Eaid, M., S. Keating, and K. Innanen, 2020b, Multiparameter seismic elastic full-waveform inversion with combined geophone and shaped fiber-optic cable data: Geophysics, 85, no. 6, R537–R552, doi: 10.1190/geo2020-0170.1.

Eaid, M., J. Li, and K. Innanen, 2018, Modeling the response of shaped DAS fibers to microseismic moment tensor sources: 88th Annual International Meeting, SEG, Expanded Abstracts, 4698–4702, doi: 10.1190/segam2018-2998378.1.

Earle, S., 2019, Physical Geology, 2nd Ed., Chapter 9.2 The temperature of the earth's interior: Creative Commons Attribution, https://opentextbc.ca/physicalgeology2ed/, accessed 13 June 2022.

Egorov, A., J. Correa, A. Bona, R. Pevzner, K. Tertyshnikov, S. Glubokovskikh, V. Puzyrev, and B. Gurevich, 2018, Elastic full-waveform inversion of vertical seismic profile data acquired with distributed acoustic sensors: Geophysics, 83, no. 3, R273–R281, doi: 10.1190/geo2017-0718.1.

Ellmauthaler, A., M. LeBlanc, M. E. Willis, J. Maida, and G. Wilson, 2020a, Real-time DAS VSP acquisition and processing: First EAGE Workshop on Fibre Optic Sensing, doi: 10.3997/2214-4609.202030013.

Ellmauthaler, A., B. Seabrook, G. Wilson, J. Maida, J. Bush, M. LeBlanc, J. Dupree, and M. Uribe, 2020b, Distributed acoustic sensing of subsea wells: The Leading Edge, 39, no. 11, 801–807, doi: 10.1190/tle39110801.1.

Ellmauthaler, A., W. Palacios, M. LeBlanc, M. E. Willis, and G. Knapo, 2019, Vertical seismic profiling via coiled tubing-conveyed distributed acoustic sensing: SPE/ICoTA Well Intervention Conference and Exhibition, SPE-194244, doi: 10.2118/194244-MS.

Ellmauthaler, A., S. Shaw, M. LeBlanc, M. E. Willis, and X. Wu, 2018, Novel retrievable FiberVSP data acquisition system with integrated gamma ray logging tool for geophone-equivalent depth accuracy: U.S. Patent Pending.

Ellmauthaler, A., M. E. Willis, X. Wu, and M. LeBlanc, 2017, Noise sources in fiber-optic distributed acoustic sensing VSP data: 79th Conference and Exhibition, EAGE, Extended Abstracts, doi: 10.3997/2214-4609.201700515.

Ellmauthaler, A., D. Barfoot, M. E. Willis, X. Wu, C. Erdemir, O. Barrios-Lopez, D. Quinn, and S. Shaw, 2016a, Depth calibration for DAS VSP – lessons learned from two field trials: 86th Annual International Meeting, SEG, Expanded Abstracts, 632–636, doi: 10.1190/segam2016-13951487.1.

Ellmauthaler, A., M. Willis, X. Wu, and D. Barfoot, 2016b, Factors affecting the quality of DAS VSP data: SEG Distributed Acoustic Sensing Workshop.

Erdemir, C., M. E. Willis, and A. Padhi, 2019, Creating 3C distributed acoustic sensing data: U.S. Patent 10,281,606 B2.

Fang, G., Y. E. Li, Y. Du, J. Ma, and D. Yu, 2018, Near-surface monitoring enabled by distributed acoustic sensing: An example of the Stanford Array Data: 88th Annual International Meeting, SEG, Expanded Abstracts, 2677–2681, doi: 10.1190/segam2018-2998119.1.

FOSA, 2022, Distributed sensing measures vibration around the world from Tonga volcano (Hifi): www.fiberopticsensing.org/p/cm/ld/fid=735&tid=309&sid=3580, accessed 13 June 2022.

Freifeld, B., R. Pevzner, S. Dou, J. Correa, T. Daley, M. Robertson, K. Tertyshnikov, T. Wood, J. Ajo-Franklin, M. Urosevic, and B. Gurevich, 2016, The CO2CRC Otway Project deployment of a distributed acoustic sensing network coupled with permanent rotary sources: 78th Conference and Exhibition, EAGE, Extended Abstracts, doi: 10.3997/2214-4609.201600577.

Frignet, B., and A. Hartog, 2014, Optical vertical seismic profile on wireline cable: SPWLA 55th Annual Logging Symposium, https://onepetro.org/SPWLAALS/proceedings-abstract/SPWLA14/All-SPWLA14/SPWLA-2014-FFFF/28354

Fung, Y., 1977, A First Course in Continuum Mechanics, 2nd ed.: Prentice-Hall.

Glaser, D., R. D. Costley, K. Hathaway, G. Boitnott, and J. Weale, 2021, Snow coupled distributed acoustic sensing for intrusion detection of polar bears in arctic camps: Symposium on the Application of Geophysics to Engineering and Environmental Problems Proceedings, 262, doi: 10.4133/sageep.33-143.

Granville, J., and M. E. Willis, 2021, Depth and distance profiling with fiber optic cables and fluid hammer: U.S. Patent Application 20210032979.

Green, A., 2021, Game-changing technology for low cost, high quality borehole seismic data acquisition: 82nd Conference and Exhibition, EAGE, Extended Abstracts, doi: 10.3997/2214-4609.202113047.

Guenot, P., 2003, Propriétés d'atténuation des fibers optiques, proceeds d'élaboration: Hermès Editions, second chapter of Fibers Optiques et Télécommunications, J. P. Meunier, ed.

Hall, K., K. Innanen, and D. Lawton, 2020, Comparison of multi-component seismic data to fibre-optic (DAS) data: 90th Annual International Meeting, SEG, Expanded Abstracts, 525–529, doi: 10.1190/segam2020-3427754.1.

Hartog, A. H., 2017, An introduction to distributed optical fibre sensors: CRC Press, doi: 10.1201/9781315119014.

Hartog, A., L. Liokumovich, N. Ushakov, O. Kotov, T. Dean, T. Cuny, and A. Constantinou, 2016, The use of multi-frequency acquisition to significantly improve the quality of fibre-optic distributed vibration sensing: 78th Conference and Exhibition, EAGE, Extended Abstracts, doi: 10.3997/2214-4609.201600685.

He, R., B. Hornby, and G. Schuster, 2006, 3D wave-equation interferometric migration of VSP multiples: 76th Annual International Meeting, SEG, Expanded Abstracts, 3442–3446, doi: 10.1190/1.2370249.

Hecht, E., 1998, Optics, 3rd ed.: Addison-Wesley.

Higginson, P., D. Purkis, and M. Webster, 2017, Advancing the application of downhole fiber optics with a novel low cost disposable deployment method: SPE Offshore Europe Conference and Exhibition, SPE-186124-MS, doi: 10.2118/186124-MS.

Hill, S. and A. Ruger, 2015, Chapter 10: Kirchhoff prestack depth migration, *in* Illustrated Seismic Processing Vol 1: Imaging, SEG, doi: 10.1190/1.9781560803621.ch10.

Hill, S. and A. Ruger, 2015, Chapter 10: Kirchhoff prestack depth migration, in Illustrated Seismic Processing Vol 1: Imaging, SEG, doi: 10.1190/1.9781560803621.ch10.

Hons, M., R. Stewart, D. Lawton, and M. Bertram, 2007, Ground motion through geophones and MEMS accelerometers: Sensor comparison in theory, modeling and field data: 77th Annual International Meeting, SEG, Expanded Abstracts, 11–15, doi: 10.1190/1.2792372.

Hons, M. S., 2008, Seismic sensing: comparison of geophones and accelerometers using laboratory and field data: MSc Thesis, University of Calgary.

Hornman, J. C., 2017, Field trial of seismic recording using distributed acoustic sensing with broadside sensitive fibre-optic cables: Geophysical Prospecting, 65, no. 1, 35–46, doi: 10.1111/1365-2478.12358.

Hornman, K., B. Kuvshinov, P. Zwartjes, and A. Franzen, 2013, Field trial of a broadside-sensitive distributed acoustic sensing cable for surface seismic: 75th Conference and Exhibition, EAGE, Extended Abstracts, doi: 10.3997/2214-4609.20130383.

Hou, Q., D. Liu, X. Wang, and W. Chen, 2021, Adaptive DAS coupling noise suppression based on local MCA: First International Meeting for Applied Geoscience and Energy, SEG/AAPG, Expanded Abstracts, 2979–2983, doi: 10.1190/segam2021-3594559.1.

Hull, R., R. Meek, H. Bello, and D. Miller, 2017, Case history of DAS fiber-based microseismic and strain data, monitoring horizontal hydraulic stimulations using various tools to highlight physical deformation processes (Part A): URTeC 2695282-MS, doi: 10.15530/URTEC-2017-2695282.

Ikelle, L., 2010, Coding and Decoding: Seismic Data, The concept of multishooting: *in* K. Helbig and S. Treitel, eds., Handbook of Geophysical Exploration, Seismic Exploration, 39: Elsevier.

Innanen, K., D. C. Lawton, K. W. Hall, K. Bertram, M. Bertram, and H. Bland, 2019, Design and deployment of a prototype multicomponent distributed acoustic sensing loop array: 89th Annual International Meeting, SEG, Expanded Abstracts, 953–957, doi: 10.1190/segam2019-3216304.1.

Innanen, K., 2017a, Parameterization of a helical DAS fibre wound about an arbitrarily curved cable axis: 79th Conference and Exhibition, EAGE, Extended Abstracts, doi: 10.3997/2214-4609.201701202.

Innanen, K., 2017b, Determination of seismic-tensor strain from helical wound cable-distributed acoustic sensing cable with arbitrary and nested-helix winds: 87th Annual International Meeting, SEG, Expanded Abstracts, 926–930, doi: 10.1190/segam2017-17664060.1.

Jaing, T., G. Zhan, T. Hance, S. Sugianto, S. Soulas, and E. Kjos, 2016, Valhall dual-well 3D DAS VSP field trial and imaging for active wells: 86th Annual International Meeting, SEG, Expanded Abstracts, 5582–5586, doi: 10.1190/segam2016-13871754.1.

Jakkampudi, S., J. Shen, W. Li, A. Dev, T. Zhu, and E. Martin, 2020, Footstep detection in urban seismic data with a convolutional neural network: The Leading Edge, 39, no. 9, 654–660, doi: 10.1190/tle39090654.1.

Jay, J. A., 2010, An overview of macrobending and microbending of optical fibers: WP1212, https://www.corning.com/media/worldwide/coc/documents/Fiber/white-paper/WP1212.pdf, accessed 13 June 2022.

Jestin, C., G. Calbris, V. Lanticq, M. Roed, and C. Ringstad, 2020, On the impact of gauge length value on distributed acoustic sensing data quality: EAGE 2020 Annual Conference & Exhibition Online, doi: 10.3997/2214-4609.202011724.

Jin, G., and B. Roy, 2017, Hydraulic-fracture geometry characterization using low-frequency DAS signal: The Leading Edge, 36, no. 12, 975–980, doi: 10.1190/tle36120975.1.

Jousset, P., T. Reinsch, T. Ryberg, H. Blanck, A. Clarke, R. Aghayev, G. P. Hersir, J. Henninges, M. Weber, and C. M. Krawczyk, 2018, Dynamic strain determination using fibre-optic cables allows imaging of seismological and structural features: Nature Communications, 9, no. 1, 2509, doi: 10.1038/s41467-018-04860-y.

Karrenbach, M., V. Yartsev, S. Cole, and R. Hooper, 2021, Long-range seismicity monitoring with DAS: First International Meeting for Applied Geoscience and Energy, SEG/AAPG, Expanded Abstracts, 1216–1220, doi: 10.1190/segam2021-3584075.1.

Karrenbach, M., S. Cole, A. Ridge, K. Boone, D. Kahn, J. Rich, K. Silver, and D. Langton, 2019, Fiber-optic distributed acoustic sensing of microseismicity, strain and temperature during hydraulic fracturing: Geophysics, 84, no. 1, D11–D23, doi: 10.1190/geo2017-0396.1.

Kiyashchenko, D., A. Mateeva, Y. Duan, D. Johnson, J. Pugh, A. Geisslinger, and J. Lopez, 2020, Frequent 4D monitoring with DAS 3D VSP in deep water to reveal injected water-sweep dynamics: The Leading Edge, 39, no. 7, 471–479, doi: 10.1190/tle39070471.1.

Kobayashi, Y., Y. Uematsu, S. Mochiji, and Z. Xue, 2020, A field experiment of walkaway distributed acoustic sensing vertical seismic profile in a deep and deviated onshore well in Japan using a fibre optic cable deployed inside coiled tubing: Geophysical Prospecting, 68, no. 2, 501–520, doi: 10.1111/1365-2478.12863.

Kuvshinov, B., 2016, Interaction of helically wound fibre-optic cables with plane seismic waves: Geophysical Prospecting, 64, no. 3, 671–688, doi: 10.1111/1365-2478.12303.

LeBlanc, M., A. Ellmauthaler, M. E. Willis, and X. Wu, 2018, Use of long gauge length distributed acoustic sensing for improved spatial resolution of first break detection in vertical seismic profiling: 80th Conference and Exhibition, EAGE, Extended Abstracts, doi: 10.3997/2214-4609.201801501.

Lindsay, N., H. Rademacher, and J. Ajo-Franklin, 2020, On the broadband instrument response of fiber-optic DAS arrays: Journal of Geophysical Research: Solid Earth, 125, 2, doi: 10.1029/2019JB018145.

Lumens, P., A. Franzen, K. Hornman, S. Grandi, G. Hemink, B. Kuvshinov, J. La Follett, B. Wyker, and P. Zwartjes, 2013, Cable development for distributed geophysical sensing with a field trial in surface seismic: Fifth European Workshop on Optical Sensors, doi: 10.1117/12.2025693.

Maida, J. L., 2008, Method and apparatus for providing hydrogen diffusion barrier for fiber optic cables in hostile environments: U.S. Patent 7,400,803.

Martin, E., 2021, Surface DAS and environmental applications: First International Meeting for Applied Geoscience and Energy, SEG/AAPG, Special Session: An introduction to DAS: Using fiber optics for geoscience applications.

Martin, E. R., C. M. Castillo, S. Cole, P. S. Sawasdee, S. Yuan, R. Clapp, M. Karrenbach, and B. L. Biondi, 2017, Seismic monitoring leveraging existing telecom infrastructure at the SDASA: Active, passive, and ambient-noise analysis: The Leading Edge, 36, no. 12, 1025–1031, doi: 10.1190/tle36121025.1.

Martuganova, E., M. Stiller, K. Bauer, J. Henninges, and C. Krawczyk, 2021, Cable reverberations during wireline distributed acoustic sensing measurements: Their nature and methods for elimination: Geophysical Prospecting, 69, no. 5, 1034–1054, doi: 10.1111/1365-2478.13090.

Martuganova, E., J. Henninges, M. Stiller, K. Bauer, B. Norden, C. Krawczyk, and E. Huenges, 2019, DAS-VSP measurements using wireline logging cable at the Groß Schönebeck Geothermal Research Site, NE German Basin: 81st Conference and Exhibition, EAGE, Extended Abstracts, doi: 10.3997/2214-4609.201901246.

Mateeva, A., J. Lopez, H. Potters, J. Mestayer, B. Cox, D. Kiyashchenko, P. Wills, S. Grandi, K. Hornman, B. Kuvshinov, W. Berlang, A. Yang, and R. Detomo, 2014, Distributed acoustic sensing for reservoir monitoring with vertical seismic profiling: Geophysical Prospecting, 62, no. 4, 679–692, doi: 10.1111/1365-2478.12116.

Merry, A., V. Lesnikov, R. Jones, A. Mitra, and P. Rennison, 2021, A DAS VSP pilot survey for 3D imaging and 4D monitoring at the Culzean Field, UKCS: 82nd EAGE Annual Conference & Exhibition, doi: 10.3997/2214-4609.202112653.

Mitrica, D., 2019, World's first earthquake detector was invented 2000 years ago in China: ZME Science, https://www.zmescience.com/science/geology/worlds-first-seismoscope-53454, accessed 13 June 2022.

Mizuno, T., S. Leaney, J. L. Calvez, F. Naseer, and M. Khaitan, 2019, The significance of gauge length in particle velocity estimation from DAS data: VSP and microseismic cases: 89th Annual International Meeting, SEG, Expanded Abstracts, 4869–4873, doi: 10.1190/segam2019-3215250.1.

Mjehovich, J., and G. Jin, 2021, Rapid surface-deployment of DAS cable for earthquake hazard assessment: AGU Fall Meeting, https://agu.confex.com/agu/fm21/meetingapp.cgi/Paper/841857, accessed 13 June 2022.

Monsegny, J., K. Hall, D. Trad, and D. Lawton, 2021, Least squares DAS to geophone transform: First International Meeting for Applied Geoscience and Energy, SEG/AAPG, Expanded Abstracts, 488–492, doi: 10.1190/segam2021-3594961.1.

Morse, P., and G. Hildebrandt, 1989, Ground-roll suppression by the stackarray: Geophysics, 54, no. 3, 290–301, doi: 10.1190/1.1442654.

Ning, I., and P. Sava, 2018a, High resolution multi-component distributed acoustic sensing: Geophysical Prospecting, 66, no. 6, 1111–1122, doi: 10.1111/1365-2478.12634.

Ning, I., and P. Sava, 2018b, Multicomponent distributed acoustic sensing: Concept and theory: Geophysics, 83, no. 2, P1–P8, doi: 10.1190/geo2017-0327.1.

Ning, I., and P. Sava, 2016, Multicomponent distributed acoustic sensing: 86th Annual International Meeting, SEG, Expanded Abstracts, 5597–5602, doi: 10.1190/segam2016-13952981.1.

Officer, C., 1958, Introduction to the theory of sound transmission: McGraw Hill Books.

Olofsson, B., and A. Martinez, 2017, Validation of DAS data integrity against standard geophones – DAS field test at Aquistore site: The Leading Edge, 36, no. 12, 981–986, doi: 10.1190/tle36120981.1.

Oppenheim, A., and R. Schafer, 1975, Sampling of continuous time signals, *in* Digital Signal Processing: Prentice Hall International.

Padhi, A., and M. E. Willis, 2021, Wavelet transform-based coherent noise reduction in distributed acoustic sensing: U.S. Patent 11,194,070.

Padhi, A., M. E. Willis, X. Wu, and A. Ellmauther, 2021, Method to improve DAS channel location accuracy using global inversion: U.S. Patent 11,073,629.

Paillet, F., and J. E. White, 1982, Acoustic modes of propagation in the borehole and their relationship to rock properties: Geophysics, 47, no. 8, 1215–1228, doi: 10.1190/1.1441384.

Podgornova, O., S. Leany, S. Zeroug, and L. Liang, 2017, On full-waveform modeling and inversion of fiber-optic VSP data: 87th Annual International Meeting, SEG, Expanded Abstracts, 6039–6043, doi: 10.1190/segam2017-17652912.1.

Poletto, F., D. Finfer, and P. Corubolo, 2015, Broadside wavefields in horizontally helically-wound optical fiber and hydrophone streamer: 85th Annual International Meeting, SEG, Expanded Abstracts, 95–99, doi: 10.1190/segam2015-5896445.1.

Saxton, L., M. Ball, S. Soulas, G. Zhan, and M. Webster, 2018, Results from a Clair DAS VSP pilot and lessons learned for DAS surveillance on Clair Ridge: 80th EAGE Conference and Exhibition, 2018 Workshop Programme, doi: 10.3997/2214-4609.201801918.

Schilke, S., D. Donno, A. Hartog, and H. Chauris, 2017, DAS and its coupling for VSP applications using wireline cable, EAGE/DGG Workshop on Fibre Optic Technology in Geophysics, doi: 10.3997/2214-4609.201700158.

SEAFOM, 2018, DAS Parameter Definitions and Tests, Measuring Sensor Performance Document – 02 (SEAFOM MSP-02).

Sloan, S. D., R. Miller, and D. Steeples, 2021, A history of tunnels and using active seismic methods to find them: Geophysics, 86, no. 3, WA49–WA57, doi: 10.1190/geo2020-0959.1.

Spackman, T., 2019, Novel orbital seismic sources at a CO_2 storage monitoring site, Newell County, Alberta: Master Thesis, University of Calgary.

Spikes, K., N. Tisato, T. Hess, and J. Holt, 2019, Comparison of geophone and surface-deployed distributed acoustic sensing seismic data: Geophysics, 84, no. 2, A25–A29, doi: 10.1190/geo2018-0528.1.

Stajanca, P., S. Chruscicki, T. Homann, S. Seifert, D. Schmidt, and A. Habib, 2018, Detection of leak-induced pipeline vibrations using fiber-optic distributed acoustic sensing: MDPI Sensors 2018, 18, 2841, doi: 10.3390/s18092841.

Tertyshnikov, K., and R. Pevzner, 2020, 3C VSP with DAS – an obvious approach: First EAGE Workshop on Fibre Optic Sensing, doi: 10.3997/2214-4609.202030016.

Thévenaz, L., 2011, Advanced Fiber Optics: CRC Press, doi: 10.1201/b16404.

Titov, A., G. Jin, Y. Fan, A. Tura, K. Kutun, and J. Miskimins, 2020, Distributed fiber-optic sensing based production logging investigation: Flowloop experiments: First EAGE Workshop on Fibre Optic Sensing, doi: 10.3997/2214-4609.202030035.

Titov, A., Y. Fan, K. Kutun, and G. Jin, 2022, Distributed acoustic sensing (DAS) response of rising Taylor 2 bubbles in slug flow: MDPI Sensors, doi: 10.3390/s22031266.

Tribolet, J., 1977, A new phase unwrapping algorithm: IEEE Transactions on Acoustics, Speech, and Signal Processing, 25, no. 2, 170–177, doi: 10.1109/TASSP.1977.1162923.

Verizon News Archives, 1997, World's longest undersea fiber cable commences Europe-Middle East-Asia operations: https://www.verizon.com/about/news/press-releases/worlds-longest-undersea-fiber-cable-commences-europe-middle-east-asia-operations, accessed 13 June 2022.

Verliac, M., V. Lesnikov, and C. Euriat, 2015, The Rousse-1 DAS VSP experiment – Observations and comparisons from various optical acquisition systems: 85th Annual

International Meeting, SEG, Expanded Abstracts, 5534–5538, doi: 10.1190/segam2015-5886544.1.

Westbrook, P., K. Feder, T. Kremp, E. Monberg, H. Wu, B. Zhu, L. Huang, D. Simoff, S. Shenk, V. Handerek, M. Karimi, A. Nkansah, and A. Yau, 2020, Enhanced optical fiber for distributed acoustic sensing beyond the limits of Rayleigh backscattering: iScience, 23, no. 6, 101137, doi: 10.1016/j.isci.2020.101137.

Westbrook, P. S., K. S. Feder, R. M. Ortiz, T. Kremp, E. M. Monberg, H. Wu, D. A. Simoff, and S. Shenk, 2017, Kilometer length, low loss enhanced back scattering fiber for distributed sensing: 25th Optical Fiber Sensors Conference, doi: 10.1109/OFS.2017.7961121.

Wilks, M., and A. Wuestefeld, 2018, Synthetic modelling of a DAS cable's sensitivity and performance in microseismic monitoring applications: Seventh EAGE Workshop on Passive Seismic, doi: 10.3997/2214-4609.201800065.

Willis, M., A. Ellmauthaler, X. Wu, and M. LeBlanc, 2022, Important aspects of acquiring distributed acoustic sensing (DAS) data for geoscientists, *in* Y. Li, M. Karrenbach, and J. Ajo-Franklin, eds., Distributed acoustic sensing in geophysics: Methods and applications, Geophysical Monograph Series, AGU and Wiley Press.

Willis, M. E., 2022a, A_Quiz_MarkEWillis_SEG_DISC_2022, GitHub repository, doi: 10.5281/zenodo.6824405.

Willis, M. E., 2022b, B_Vz_to_Strain_Rate_MarkEWillis_SEG_DISC_2022, GitHub repository, doi: 10.5281/zenodo.6824412.

Willis, M. E., 2022c, C_Strain_Rate_to_VZ_MarkEWillis_SEG_DISC_2022, GitHub repository, doi: 10.5281/zenodo.6824417.

Willis, M. E., 2022d, D_Phase_to_Geophone_MarkEWillis_SEG_DISC_2022, GitHub repository, doi: 10.5281/zenodo.6824615.

Willis, M. E., 2022e, ExploreDAS, GitHub repository, doi: 10.5281/zenodo.6828738.

Willis, M. E., A. Ellmauthaler, G. Wilson, and O. Barrios, 2021, Subsea fiber topology and effect on 3D and 4D subsea DAS VSP images: EAGE GeoTech 2021 Second EAGE Workshop on Distributed Fibre Optic Sensing, doi: 10.3997/2214-4609.202131038.

Willis, M. E., A. Ellmauthaler, O. Barrios, G. Wilson, and X. Wu, 2020a, The requirement for optical noise reduction on subsea DAS VSP systems: SEG Kuala Lumpur Workshop Exploring in Presence of Complex Overburden.

Willis, M. E., W. Palacios, A. Ellmauthaler, and X. Zhao, 2020b, Mitigation of zigzag noise on DAS VSP records acquired in vertical wells: EAGE 2020 Annual Conference & Exhibition, doi: 10.3997/2214-4609.202011089.

Willis, M. E., O. Barrios, V. Leung, and A. Ellmauthaler, 2019a, Noise removal for distributed acoustic sensing data: U.S. Patent 10,401,519.

Willis, M. E., X. Wu, W. Palacios, and A. Ellmauthaler, 2019b, Understanding cable coupling artifacts in wireline-deployed DAS VSP data: 89th Annual International Meeting, SEG, Expanded Abstracts, 5310–5314, doi: 10.1190/segam2019-3208294.1.

Willis, M. E., A. Ellmauthaler, M. LeBlanc, W. Palacios, and X. Wu, 2018, Comparing distributed acoustic sensing, vertical seismic profile data acquired with single- and multimode fiber optic cables: 88th Annual International Meeting, SEG, Expanded Abstracts, 4674–4678, doi: 10.1190/segam2018-2996212.1.

Willis, M. E., A. Ellmauthaler, X. Wu, D. Barfoot, C. Erdemir, O. Barrios-Lopez, D. Quinn, and S. Shaw, 2016a, Calibration of distributed acoustic sensing (DAS): 78th Conference and Exhibition, EAGE, Extended Abstracts, doi: 10.3997/2214-4609.201601599.

Willis, M. E., C. Erdemir, A. Ellmauthaler, O. Barrios, and D. Barfoot, 2016b, Comparing DAS and geophone zero-offset VSP data sets side-by-side: CSEG Recorder, 41, no. 6, https://csegrecorder.com/articles/view/comparing-das-and-geophone-zero-offset-vsp-data-sets-side-by-side, accessed 13 June 2022.

Willis, M. E., A. Padhi, and D. Barfoot, 2014, The value of many DAS VSP channels and the effect of gauge length on signal quality: SEG DAS Workshop.

Willis, M. E., D. Barfoot, A. Ellmauthaler, X. Wu, O. Barrios, C. Erdemir, S. Shaw, and D. Quinn, 2016c, Quantitative quality of distributed acoustic sensing vertical seismic profile data, The Leading Edge, 35, no. 7, 605–609, doi: 10.1190/tle35070605.

Willis, M., R. Foy, A. Padhi, and O., Barrios, 2015, A quick, flexible, layer-based VSP modeling algorithm, Geophysical Society of Houston presentation, 17 March 2015.

Wilson, G., M. E. Willis, and A. Ellmauthaler, 2021, Evaluating 3D and 4D DAS VSP image quality of subsea carbon storage: The Leading Edge, 40, no. 4, 261–266, doi: 10.1190/tle40040261.1.

Wilson, G., X. Wu, A. Ellmauthaler, and M. E. Willis, 2020a, Real-time processing and control of DAS VSP surveys: U.S. Patent Application 20200271811.

Wilson, G., A. Ellmauthaler, B. Seabrook, J. Maida, J. Bush, M. LeBlanc, J. Dupree, and M. Uribe, 2020b, Topside distributed acoustic sensing in subsea developments: 90th Annual International Meeting, SEG, Expanded Abstracts, 3372–3376, doi: 10.1190/segam2020-3428141.1.

Wilson, G., A. Ellmauthaler, B. Seabrook, J. Maida, J. Bush, M. LeBlanc, J. Dupree, and M. Uribe, 2020b, Topside distributed acoustic sensing in subsea developments: 90th Annual International Meeting, SEG, Expanded Abstracts, 3372–3376, doi: 10.1190/segam2020-3428141.1.

World Oil, 2008, Are wells perforated on depth today? https://www.worldoil.com/magazine/2008/january-2008/features/are-wells-perforated-on-depth-today/, accessed 13 June 2022.

Wu, H., K. S. Feder, A. A. Stolov, S. D. Shenk, E. M. Monberg, and D. A. Simoff, 2020, High-temperature enhanced Rayleigh scattering optical fiber sensor for borehole applications: Optical Components and Materials, doi: 10.1117/12.2538120.

Wu, X., M. E. Willis, and D. Barfoot, 2019, Gauge length effect and gauge length conversion: U.S. Patent Application 2019/0331819 A1.

Wuestefeld, A., 2021, Keynote: Corkscrew well paths for improved microseismic event locations with DAS recordings: EAGE GeoTech 2021 First EAGE Workshop on Induced Seismicity, doi: 10.3997/2214-4609.202131056.

Wuestefeld, A., and M. Wilks, 2019, How to twist and turn a fiber: Performance modeling for optical DAS acquisitions: The Leading Edge, 38, no. 3, 226–231, doi: 10.1190/tle38030226.1.

Yang, J., J. Shragge, and G. Jin, 2021, 4D DAS fiber-coupling effects in freezing near-surface ground conditions: First International Meeting for Applied Geoscience and Energy, SEG/AAPG, Expanded Abstracts, doi: 10.1190/segam2021-3582508.1.A

Yu, C., Z. Zhan, N. Lindsey, J. Ajo-Franklin, and M. Robertson, 2019, The potential of DAS in teleseismic studies: Insights from the Goldstone experiment: Geophysical Research Letters, 46, no. 3, 1320–1328, doi: 10.1029/2018GL081195.

Yu, G., Z. Cai, Y. Chen, X. Wang, Q. Zhang, Y. Li, Y. Wang, C. Liu, B. Zhao, and J. Greer, 2018, Borehole seismic survey using multimode optical fibers in a hybrid wireline: Measurement, 125, 694–703, doi: 10.1016/j.measurement.2018.04.058.

Zhan, G., and J. Nahm, 2020, Multi-well 3D DAS VSPs: Illumination and imaging beyond the wellbores: 90th Annual International Meeting, SEG, Expanded Abstracts, 3798–3802, doi: 10.1190/segam2020-3426032.1.

Index

D

E

F

G

H

I

J

K

L

M

N

O

P

Q

R

S

T

U

V

W

Z